Stairway to the Smartphone

- My Life -

My Invention

By

Dr. Rocco Leonard Martino

Published By

BlueNose
PRESS, INC.

Printed in the United States of America
Published June 2019

ISBN: 1-7324590-2-9
ISBN-13: 978-1-7324590-2-1

Cover Design: Joseph A. Martino
Interior Layout Design: Joseph A. Martino

For more information on this title please visit:
www.PublishWithJAM.com

DEDICATION

This book is dedicated to my son Joseph Martino and to my loyal associate Joseph Looby. Without these two, this book would never have been written, and the true story of the CyberFone might never have been told. It is a wonderful story, full of surprises. It was an idea, that became reality, and which finally captured the public interest when a marketing genius with an unlimited advertising budget saw its potential. Apple became the dominant force in the new industry. It became the most valuable company in the world, and it now has a nest egg of over $225billion in cash, while I, the inventor, am still in debt from the patent and legal fees to license patents which were ignored and pushed aside by rather dubious business practices.

ACKNOWLEDGEMENTS

I am indebted to my son Joseph Martino who edited this book and helped me in other ways to get this book written and ready for publication. He, in particular, was instrumental in putting together the Appendices. I also appreciate all the hard work done by Christine Monigle who transcribed this text from my dictation, and to George Ingram and Jack Schrems for meticulously proofreading this book and offering their suggestions for the text.

Most of all I am indebted to my wife Barbara who, among many other things she does for me, now drives me everywhere with great patience.

I am also indebted to Joseph Looby. For years he worked with me in the CyberFone companies in an effort to have people listen to what we were saying and doing.

What I did was to bring the world to the palm of everyone's hand. My invention changed the world forever.

REVIEWS

Stairway to the Smartphone - My Life - My Invention will encourage those of us who read this life's work to want to enhance technology as well.

All will find, in addition, an incentive in these pages to imitate Dr. Martino. Expressing through invention one's opinions about advancing the technologies of the day.
- *Rev. George W. Bur, SJ, Superior of the Jesuit Center in Wernersville, PA*

In the movie, *Rocky*, the title character loses the championship fight but is heralded for his grit and determination. Rocky Martino lost his fight with the patent lawyers, preventing him from collecting royalties due him from his invention of the Smartphone. But just like his name sake in the Oscar-winning movie, Dr. Martino continues to be an inspiration for his fighting spirit and for his successes in his other business ventures and in life.

Stairway to the Smartphone - My Life - My Invention chronicles Rocky's life and his experiences with NASA, his ground-breaking achievements in the world of financial information systems and his time serving the Vatican. He talks about the teachers and mentors who helped him along the way and gives credit for his success to his loving wife, Barbara and his family.

Although Dr. Martino has endured many health issues, his recollection of events are presented with stunning detail. This book is a must read for any budding entrepreneur looking to go the distance with their inventions and ideas. One can only hope that they finish on top - just like Rocky Martino!

*- **Tom Burgoyne, The Best Friend of the Phillie Phanatic and Co-Author of Pheel the Love!***

Stairway to the Smartphone - My Life - My Invention is a fascinating and educational story from the early days of a 90-year-old man - who "came within a whisker of being the richest man in the world" with his invention of the CyberFone, which has come to be known by the world as the Smartphone - to his remarkable acceptance of the impending end days.

Dr. Martino honestly admits he wrote the latest of his books as proof of his assertion that he invented the product that everyone of us holds in our hands more than holding hands now and certainly can't live without.

He already detailed his life and lessons learned in a three-volume set of his memoirs and he continues to explain how his faith, education, family and friends shaped his fascinating and accomplished life before telling how modifications in the patent law prevented him from receiving any monetary compensation for his latest invention. Although Dr. Martino didn't receive riches from inventing the Smartphone, no one can take away his admirable journey to get there and his rich life.

- Kevin Callahan, Sports Writer and Author of The Black Rose, The Fish Finder, and The Chess Game

Stairway to the Smartphone - My Life - My Invention is the extraordinary story of one man's struggle to gain worldwide recognition for his invention of the Smartphone, one of the technological breakthroughs of our time. But it is also the narrative of a life devoted to intellectual curiosity, innovation, and--most important--to his family, his many friends, and his faith.

- George Ingram, Freelance Writer and Co-Author of Jersey Lawman: A Life On the Right Side of Crime

Stairway to the Smartphone - My Life - My Invention is an expansion of Doctor Martino's life on the frontier of technological advancements in the world, from the invention of reentry heatshields for manned spaceflight, to the invention of computer languages, to the invention of the world's first Smartphone. His technological contributions have truly changed the world. I am glad he has shared them with us; as you will too, his reader.

- *Paul C. Peterson, P.E., Fellow of the American Society of Civil Engineers, Retired Chair of the Theology Department at Bishop Shanahan High School, Downingtown, PA*

The reader moves along throughout *Stairway to the Smartphone - My Life - My Invention* as if following the life of someone other than Rocky Martino. The many inventions, personal life, and other accomplishments that come before the Smartphone authenticate the saga of this invention.

- *John J. Schrems, PhD, Professor Emeritus of Political Science, Villanova University*

From Dr. Rocco Martino's involvement with the ENIAC to his conception and creation of the Smartphone, *Stairway to the Smartphone - My Life - My Invention* is a fascinating recount of how he helped change the world with his inventions and his great faith in family and mankind.

- *Albert W. Tegler, President & CEO, Tegler Benefits Group*

WORKS BY ROCCO LEONARD MARTINO

FICTION

The Cross of Victory

Christianity: A Criminal Investigation...

The Resurrection: A Criminal Investigation...

9-11-11: The Tenth Anniversary Attack

The Plot to Cancel Christmas

NONFICTION

Motivational Reflections – At Sunset

The Coming Technology Tsunami

Memories: Volume I - Stories for My Grandchildren

Memories: Volume II - Scientist and Writer

Memories: Volume III - Changing the World

Rocket Ships and God

People, Machines, and Politics of the Cyber Age Creation

Finding the Critical Path

Applied Operational Planning

Allocating and Scheduling Resources

Critical Path Networks

Resources Management

Dynamic Costing

Project Management

Decision Patterns

Decision Tables with Staff of MDI

Information Management

Integrated Manufacturing Systems

Management Information Systems

MIS Methodology

Personnel Management Systems

IMPACT 70s with John Gentile

TABLE OF CONTENTS

PREFACE

This book is an autobiography of the CyberFone, the very first Smartphone. The book is in two parts: chapters one through seven, where I tell my story; and the appendices, which are integral documentations of the story.

I spent almost fifteen years of my life inventing, building prototypes, and demonstrating the CyberFone only to have it stolen from me by manipulative lawyers and greedy businessmen.

My CyberFone, was a handheld smart device before others called it the Smartphone. I wrote papers about the Smart Society but didn't have the sales savvy to call the CyberFone a Smartphone. What I hope to accomplish is to set the record straight on a lot of things that I did and on a lot of things that have happened during my battle to get the CyberFone in use. Then it exploded. It was certainly an exciting, albeit a costly, chapter in my life. Infringers of my patents made hundreds of billions, while I was left with a $10million debt.

This is not a sad book. It is a realistic look back at 90 years. They were not quiet years. They were exciting years. They were years full of adventures, interesting people and interesting places. The cornerstone of these years, as I will show in the first chapter, was my marriage to a remarkable lady, Barbara D'Iorio. The story will also show how

I came within a whisker of being the richest man in the world.

Here I am now approaching my 90th birthday in June, stricken with wide-spread bone cancer from a metastasizing of prostate cancer. The bone cancer came as a total surprise. For six years I had suffered from high-grade invasive bladder cancer. It was only a question of time before that particular cancer would kill me. And yet after six years and 13 surgical procedures, it suddenly disappeared. I am now free of bladder cancer. This of course blind-sided me and the doctors. My PSA reading for years had been less than one. Medicare refuses to pay for PSA tests after the age of 80. But when I fell and broke my six ribs, medical tests included the PSA. What a surprise to everyone when that turned out to be 284. This immediately led to a bone scan which showed extensive bone cancer in my spine, in my legs, and in certain portions of my skull. Suddenly I had to face the prospect that I had a new foe that was attempting to kill me. I had been riding high and looking forward to my 100th birthday, feeling exuberant as I was approaching the Milestone 90. Suddenly I was stricken with this new need to appraise my mortality.

So, there it is. What has happened in the short period since the PSA discovery, is that I have been on a regime of Zytiga pills. These are rather expensive, running to a total of about $9,000 -

$10,000 per month. I paid a substantial part of this the first month and am now paying about $600 per month. These together with periodic injections of Leuprolide Acetate, a hormone for prostate cancer, or its trade name Lupron. My PSA dropped to 12.5 when tested in April. We hope to get it down to one or less. The theory is to knock out the ability of the prostate cancer to spawn the spread of cancer cells and in this fashion to wean, eliminate, or reduce the cancer in the bones.

That's the theory. Now with a lot of prayer we hope it will happen.

I want to get my story out while I am still functional. That's the reason for this book. Hopefully this is not my final book, but I am writing while I still can. This book will be an assessment of the great adventure I lived, but specifically it will be a detailed history of my invention that changed the world forever – the CyberFone, or as it came to be known, the Smartphone.

How did I come to invent the CyberFone?

I always referred to this invention as the CyberFone. I wrote articles and spoke about the Smart Society. I only invented it! Steve Jobs called it the iPhone. The Apple advertising money got it going. They even used the camera, which I put into the CyberFone, as a means of selling the iPhone.

Steve should have called it the CyberFone. They followed the patent with everything else.

CHAPTER 1: ROOTS, MARRIAGE AND PARTNERSHIP

Why this book?

I invented the CyberFone, as I called it – the Smartphone as it is now known. I am writing this book as proof of that assertion. I am doing this to establish my legacy. We do not often think about our legacy. I never did, but now I do. I am under sentence of death from cancer. I know it's rather imminent. It could be this year. I've been told by the oncologists and cancer specialists that I will have about 5-7 years before the current cancers will kill me. That is taking into account the ongoing treatments. The initial response of my body has been quite good and very encouraging with regard to prolonging my lifespan. But suddenly I feel differently. I don't think I have 5-7 years to go. By the way I would be 95-97 by then, so something else in the end will probably kill me. Right now, I just seem weak. My mind is a sharp as ever, but I find it very difficult to do anything physical. I have a physical therapist who comes twice a week. She measures my walking ability in terms of distance. And last week I had an all-time high of four hundred feet. I used to walk at least five miles a day. Now four hundred feet leads to fatigue and rather severe pain. I see that as a sign of the impending end.

That doesn't frighten me. I have no fear of dying. I have an intense sadness. The sadness comes in thinking of my grandchildren. I just so want to see them graduate from college, I just so want to see them start on their own road for their careers and their life. I am hoping I might even be around to see more great-grandchildren. We have one now, and she is a gem, a perfect doll, a chatterbox and smart as a whip. I think her IQ is off the scale. Of course, I am biased. But anyone who spends five minutes talking with Sophia can see that this is a remarkable young lady. She is now six, going on thirty-six, and I am hoping to be around to see her in school.

That is the great regret or sadness that I face as I look at my mortality.

Barbara – Wife and Partner

The great joy of my life is my wife Barbara. She has been the strength and support and a partner in everything that we did since we were married. There is a great sadness in the thought that I will be leaving Barbara, yet I never feel that I will ever leave her. I think that no matter what the dimensions of the great divide are, I will never be apart from her. Theologically it is said that all marriages are dissolved with death. Somehow, I don't think that ours will be. I think that even though my physical life is ending, my marriage and my partnership with Barbara will never end.

Wait till I tell you how we courted.

On her 80th birthday, we celebrated with family and friends at Overbrook Golf Club. All had a great time. It was my great joy to offer a toast to her. Here it is:

> Barbara. What can I say? Fifty-seven years. I can't imagine those years without her. We have traveled the world, worked hard for the Church, helped innumerable number of people. She also worked hard for her alma maters – Little Flower High School, and Chestnut Hill College. She was on the Board (of which?) for 23 years, 4 years as Chair. In between, she bore and raised 4 boys. Finally, she was my wife – companion, confidant, nurse, and love. I can't say enough of how much I love her. I try to express my thanks, but how? She will be canonized for living with me without a single conflict. God bless her and the wonderful parents who raised her. Please join me in toasting this remarkable and good woman. Barbara!

God bless her!

Courting Barbara

It was 1960 in October that we first met at the home of John Mauchly, my partner at Mauchly Associates and the man who with J. Presper Eckert, co-invented the digital computer. They invented the ENIAC. John was the spark with the design concepts and Pres was the engineering genius who built the machine. It was said Pres could take two pieces of wire and create a circuit from them. As an 18-year-old he developed circuits for Hi-Fi music which he patented. Strangely enough the last time I saw Pres was just the week before he died. It was at a concert at Daylesford Abbey in suburban Philadelphia. He was recording it using the tremendous echo chamber of the Abbey.

Going back to Barbara, there had been a board meeting of Mauchly Associates in Ambler, Pennsylvania, on Saturday night somewhere around October 30th of 1960. It was the last Saturday of October. The following day, Sunday, John and Kathleen were napping, and the kids were playing. The doorbell rang at about 3pm and I went to answer it and there was a beautiful young lady who said, "I'm Barbara." I was told she was coming but no one told me of the devastating effect she would have on me.

I didn't exactly know what to do with her in the time before John and Kate would appear. Billy

Mauchly, who was then 8 years old and is now well into his late 60's, was there and together with him we played a game called Bridge-It. I don't know anything more than remembering the name. In any event I was quite taken with Barbara and I knew I was interested in getting to know her better.

Dinner was stimulating and with a certain amount of kidding by Kay, who was the instigator of this meeting. It all started one day when she said to me "Why aren't you married?" I told her "John keeps me traveling around the world. I am too busy." She said very quietly "I will fix that." And she did. By the way Kay has a building named for her now in Dublin, Ireland. She was one of the first women of the computer, one of the six hired by Eckert and Mauchly for the ENIAC program. John married Kay after his first wife tragically drowned in a swimming accident.

So, there we were, and we had a great time. Sydney Mauchly, John's daughter, had been a classmate of Barbara, at Chestnut Hill College. That led to the association. As a matter of fact, Barbara had been godmother to Sydney's first child, Jimmy.

After dinner, Barbara and I said our goodbyes. For the next few weeks Sydney kept bugging me if had I phoned Barbara. In addition to my normal work schedule, I had gone to Europe and to California, and I really didn't settle down until

Christmas. So immediately after the holidays, I phoned Sydney and got Barbara's phone number. It was February when I called. Barbara's father answered. I wished him a Happy New Year and asked for Barbara. He gave her the phone and said, as reported later by Barbara, "Some idiot with a foreign accent just wished me a Happy New Year." She came to the phone and I asked if she was available to go out with me since I would be in the Philadelphia area in late February. We arranged to go out the last Saturday of February. Lo and behold we were struck by a snowstorm on the entire East coast and I was snowbound in Toronto. I finally tossed in the towel and phoned her somewhere around five o'clock on Saturday and told her I couldn't make it. I asked if she could go out with me the following Saturday.

I flew to New York and spent the week there and then took the train to Philadelphia. I stayed at John Mauchly's house and borrowed a car from him. It was a Karmann Ghia, a sports version created by Volkswagen. I don't know if it's still sold. In any event I wasn't sure how to drive it. I also wasn't so sure how to get to Barbara's house, but I got there. I went up to the doorbell and opened the storm door, as a button came off my jacket which I caught it in my hand. And there I was as her mother answered, I stood there with a coat button in my hand. She started to laugh and said, "Come on in, I'll take care of it." That was the way she was, taking care of everything for us even after we got married. Barbara's mother, Esther, was a godsend. She took me into the kitchen, and we sat for a few minutes, or at least I sat with Barbara while her mother made me a cup of coffee and then sewed the button onto my jacket. Barbara explained to me that we were going to the Latin Casino to see Harry Belafonte. I said, "That's great, show me how to get there." That created a burst of laughter. In any event, we took off and got to the Latin Casino. I decided I would have some wine with dinner and ordered sparkling Burgundy. The waiter suggested that I wouldn't be able to afford it. That shook me and I didn't know what to do so I told him to serve it to me anyway. I ordered a steak for myself, and I don't remember what Barbara had. The steak must have been bad, because you could smell it as the waiter came in with it. I sent it back and ordered another one. This too came stinking to high heaven and I wasn't sure what to do. In the meantime, Barbara was laughing her head off. I ordered something else that came, and it was alright.

The final straw of merriment came when the waiter presented the bill and I gave him my Diner's Club card. The waiter said they don't accept those here. I said "What?" I had used it all over the United States and in Europe and here this waiter was telling me that he would not accept it. I just very quietly asked for the manager and said, "This card is good internationally and it says so." He said, "Yes, it is." I said, "Your waiter will not accept it". He said, "I will." That was it.

Harry Belafonte was tremendous. The show was over early, and I asked Barbara if she wanted to go somewhere for an aperitif. She said, "Sure let's go to the Cherry Hill Inn." I said, "Show me how to get there." And she did, so we had a drink there and talked. I was fascinated. Then I drove her home. I wasn't too familiar with the car and couldn't put the heater on, but she was cold, so I gave her my jacket and she said that was heraldic and noble of me. At the time she had a body temperature that was always cold, and I had a body temperature that was always hot. Somewhere in the last 10 years we have reversed that where she is a budding furnace and I'm always cold. In any event we are a compatible pair. So, we finally got home, and she asked, "Will I see you again?" and I said, "I think so, but I have to go to Europe." And that's how we parted. I don't think we kissed but I just took her to the door and her father was right there waiting. It was pretty late, it was around 2 am. In any event that was our first date, March 4th of 1961.

I did go to Europe and came back. While I was there, I kept thinking of Barbara. I bought a Cameo bracelet at the Vatican Museums for her. I phoned and we arranged to go out on a Wednesday night in March. We went to the Scioli's Restaurant in Philadelphia. I didn't remember the place, but Barbara reminded me as I wrote this. We had a wonderful time. I gave her the bracelet. When we got home, I did kiss her goodnight.

She was very excited, and she showed the bracelet to her father and he said, "Give it back." And she said, "No way!" He said, "That man is serious." And she said, "So what." And that was our second date.

I think we went out once before I had to go to California. Finally, I asked her to come to Toronto for Memorial Day holiday. And she agreed. She met me in New York, and we went from there. Pop was taken with her. As we went to Church on Sunday of that weekend, he grabbed me by the sleeve and said quietly, "Don't let her go. She is gold." He loved her instantly.

There's a certain amount of humor with the invitation to go to Toronto. I phoned my father (Pop) and asked him to please send her a written invitation. He was married at that time to Bea, his second wife, because my mother had died in 1938. He married Bea in 1950. She thought it was ridiculous for her to send a note and she wouldn't do it. Pop just told her to send a note. And she did. In any event Barbara received the proper invitation she insisted on, correctly, and we went to Toronto. We flew back to

Philadelphia and I was in the office on Monday getting ready to go to London. She phoned me and said she'd been fired. I burst out laughing. I said, "That way we can get married sooner." She didn't take it too calmly because she really enjoyed her job. But she had been fired for insubordination because she took off for the day in between Memorial Day and the last work day. She asked for the day off and her boss, who I think was actually jealous of her, refused. This was certainly not normal. In any event she was fired. I laughed my head off and said, "Fine, we'll get married." We arranged to get married in September.

So, there we were, at that time with about eight or nine dates having met in October with our first date in March and we were getting married September 2nd of 1961.

For the record, Peter was born 24 months after the wedding. I know what everybody thought. I egged them on by saying "I have to get married." Then everything went silent. I inwardly laughed. They didn't ask why. If they had, I would have told them. The reason was I couldn't afford the phone bills. I was in London on my 32nd birthday and was suddenly lonely. I phoned Barbara and it cost me $128 in 1961. That's about ten or more times in today's money. I really couldn't afford it.

It was a grand and glorious wedding, with about 160 people attending. Following the reception, we drove to New York and spent three days there before sailing to Europe on the USS Independence for our honeymoon. We spent seven weeks in Europe. In another book I have told that story, *Memories: Volume I – Stories for My Grandchildren*, published by BlueNose Press in 2015 and distributed by Amazon.

That's how I courted Barb. That's how we married. What a gem. I love her more than ever. I couldn't live without her. She has been my partner, and my love and support for over 57 years.

My Genetic Roots

I was born at home on Manning Ave in Toronto on June 25th of 1929. My mother was Josephine DiGiulio-Martino and my father Domenico Martino. My mother was a ravishing green-eyed redhead who would toss her hair and laugh and laugh and laugh. My father was a blue-eyed blond who was a master chef, an artist, and later in life Canada's foremost chef. My father lived to a good age of 82. My mother died very young, at 37. She had a five-month miscarriage in 1936 and survived for two years until she finally died at 10pm on January 25th of 1938. She died before my eyes.

We had a great party that night and Aunt Lena had come over, and the three of them had a grand time. My brother and I were allowed to stay up and join in. Aunt Lena was my mother's aunt, the youngest sister of my grandmother. Aunt Lena wasn't much older, about 10 years or so older than my mother. She lived close by and was a great friend of my mother. After Aunt Lena left, my mother went upstairs to go to bed and my father shooed me up. I followed behind her and I was heading for the bathroom across the hall from her bedroom. I saw her go in her bedroom and then I heard a thump. So, I turned and went in the bedroom and there she was on the floor. I called my father and my grandmother, who happened to be living in the house at the time. They came up to see what they could do. A doctor, Bob Miller, lived within four doors of our house. He came running over and told us there was nothing he could do. She was gone!

By happenstance, I was also in the room at 3pm on December 11th of 1982 when my father took a deep breath, sighed, and was gone.

I think I have been deeply gifted and privileged to have been with both of my parents when they died. In the first case I was only eight years old and didn't really understand, but I understood enough that I didn't cry at the funeral and didn't cry for three weeks and then came home and burst into tears and cried for at least a half hour.

I didn't cry when my father died, I just sat there with him until my brother arrived. In my mind I was still talking to him.

And so, life goes on.

What are my roots? I have outlined these in detail in my three-volume autobiography so there's little point in repeating it. These books are entitled:

- *Memories: Volume I – Stories for My Grandchildren*
- *Memories: Volume II – Scientist and Writer*
- *Memories: Volume III – Changing the World*

All were published in 2015, by BlueNose Press and distributed by Amazon.

My mother was a loving and lovely woman. My father was equally loving and caring. They were a wonderful pair and I was privileged to have them as parents. I inherited my mother's red hair and my father's blue eyes. From both I inherited a never-say-die attitude and the ability to think logically. I also inherited from both the ability to laugh.

I was also privileged to have someone who was instrumental in rearing me from the age of 10. She was my father's youngest sister, Maria, who came from Italy specifically to care for his two sons. She lived with us until she married and left, after which I was well on the way, 18 or so. My Aunt Maria was a wonderful person and once again I was very privileged to have a loving and caring person to look after me as I grew up.

My Environmental Roots

These are my genetic roots. Roots are also associated with environment and residence. I have lived in only a few places, but I have been to many places. I can safely say that I have traveled the world. I have been to Singapore, Hong Kong, Taipei, Seoul, just about every state and province in the United States and Canada, most of Europe, parts of Africa, and many of the island kingdoms of the world. From the point of view of residences, I lived in very few. In Toronto, I lived in only three places until I married. There was a certain stability of this continuity of addresses which I continued in my own life. I have lived in the same house in Philadelphia since April 1st of 1965, although we have significantly remodeled it. I have lived in the same summer home in Sea Isle City, New Jersey, since February of 1969. Once again, we made significant renovations to it.

Separately however we had an apartment in London, in the West End, in Knightsbridge, across from the Horse Guard's Parade in Hyde Park. We were a stone's throw from Royal Albert Hall. We could walk to the Brompton Oratory for Mass, including Mass in Latin. It was a short walk to the famous Harrod's Department Store.

In Rome we lived in various hotels but most especially the Minerva Hotel, a six-star hotel. We even had the same room assigned to us whenever we showed up.

In London, we always stayed at the Marriott on Grosvenor Square. It had been the International Hotel before Marriott acquired it. I still remember the late afternoon with the sudden requirement to be in London the next day. I made the air reservations and took a cab to the airport. After paying for the cab, I had two dollars in my pocket, thankfully American ones at that. In London, I took a cab to the Marriott and asked the Hall Porter to pay the bill.

Parenting

It was this element of familiarity which was very important as we travelled and when we had children. The children grew up in the same house that they now visit. We tried to give them a sense of continuity and I think this has been accomplished. I strongly recommend that house jumping, residence jumping, and city jumping is not conducive to a close-knit family life. We moved into our house when Peter was one. I still remember him walking in his blue coat and his blue cap stomping through the empty room, which is now our dining room. This is a memory that goes on forever. I remember the Christmas trees that we had, and I still remember playing, lying down, with all four boys on top of me.

There's one picture with the five of us in front of a Christmas tree that I still treasure. After all, the home is where the heart is and my heart is where my family is, and where my family was.

I still remember the bedrooms the boys occupied as they grew up, and how often I would walk through the night holding them when they were sick. I loved doing this. It gave me a sense of protecting them from all the ills of the world. It gave me a fuller realization of fatherhood.

I will always remember sitting in a rocking chair holding them, feeding them, loving them. I miss that now. That's what life is all about. They are now doing that for their own children. It gives me great solace to see that. I feel that my life is complete.

These, then, are my roots. and a more complete history can be found in Volume I of Memories, "Stories for my Grandchildren." In that book, there is a chapter about my brother Jack. In summary, no one could have a greater brother than I did.

So, then, these are my roots – genetic and environmental. The final elements of my roots come from my friendships.

CHAPTER 2: EARLY DAYS OF A WORKING CAREER

I was always inventing things in all the jobs I had. Actually, it started even before I entered college. My father had teamed up with a man called Michael Georges to create a new restaurant called the "Silver Rail." He asked me if I could design the steam table with certain new features he wanted. I was 16 at the time. I had never had a course in drafting. But I still produced a diagram with full details and how to do it. At the time, I was a high school student attending De LaSalle College in Toronto, or "Del" as we affectionately called her.

My high school years at "Del," were the happiest of my life until I met Barbara. I was only 12 when I entered Del, on a partial scholarship. The IQ tests I took on the test day were a total surprise to me. Even without preparation, I came out high enough to win the scholarship - $25 off the regular tuition of $75. Fees were relatively low in 1942. Del was experimenting with the concept of a "Gifted Students Program." We were enrolled in a Scholarship Class for Gifted Students and had to maintain Honor Ranking to stay in the group. We didn't proceed to graduate sooner or go beyond the Government mandated curriculum. We just took more courses. Everyone studied French and Latin. We also had studied German and Spanish. We read more books. We spent a great deal more time in

21

conversation in the different languages, and we had model government in our class. It was quite an experience. We also had special teachers who, at times, were considered slave drivers.

I was at Del during the war years. In fact, 1942 was the worst time for our side. Del had a Cadet Corps, affiliated with the Irish Regiment of Canada. At the age of 12, during drill I had to stand at attention and not scratch my nose when it itched. This self-discipline was very important to me as I grew older and encountered difficulties. When things were tough, I got tough and self-discipline took over for the "battle". By the way, I ended up commanding the Signal Corps of the Battalion, using every form of signaling capability in the Canadian Army then.

At that time, Ontario had five years of High School (recently it has adopted a four-year plan for high school). For us, the fifth year counted as the first year of College. Hence it was possible to graduate from the University after three years with a regular Bachelor's degree; or in four years with an Honors Bachelor Degree. I took an Honors Course in Mathematics and Physics. After graduation, I earned a Master's degree in one year. My Bachelor's degree was the equivalent of a Master's degree at Harvard or MIT. Because of this, we were readily admitted to the finest universities in the United States for graduate degrees. With my "First," an equivalent of a Summa cum Laude, I also received fellowships and scholarships.

Hard work paid off. At Del I had set academic records which I think still stand to this day. I also played on the school Football Team that won the City Championship. In Hockey, we won our Junior B league. In Baseball, I had the highest batting average.

At the conclusion of High School, Ontario required mandatory government exams. These were termed "Departmentals." Standings in these qualified you for scholarships to Canadian Universities, especially Ontario Universities, and admission to Honor Courses.

In my high school studies, Latin became a true joy. In third year, I took off and completed all the prescribed work in Latin Authors and Latin Composition for the three years that followed. I did not attend classes but worked on my own with my own schedule. I discussed problems with teachers as I went. I received special permission to do this. My Grade Average was 98.3% so it was readily given. At the end of third year, I took the fifth-year Departmental exams and scored just about the highest in the whole province. To this day, I still read Caesar's Gallic Wars and Virgil in the original Latin.

At the end of the fifth year, in June, I wrote 12 Departmental exams. One night in late August of 1947, I was reading the paper and suddenly saw the heading "Scholarship Results Announced." I skimmed the list and suddenly saw my name. I whooped. Then I saw my name again. I did a jig. The first scholarship covered four years of study at University College of the University of Toronto for any Honors Course I wished to follow. The second would cover all my texts and incidental expenses. I had won a free University Education. The scholarships stated that I had earned the highest grades in English, History, Chemistry and Physics. I enrolled in the Honors Course in Mathematics and Physics.

M&P, as we termed it, was the toughest course in the University. We began with 149 students, all of whom had straight A's in Math and Science. We graduated 21 students, seven of whom had Summa's. I with a "I,1" in my group, Mathematics. That meant a "First," Equivalent to a Summa, and First in the Group.

The news of the Scholarships made us all ecstatic. We were not poor, but certainly not rolling in money. Even with fees that would have only cost about $200 a year, paying them would have been difficult. I had summer jobs and worked weekends to help pay the bills at home. My father was still paying my mother's medical bills which took every cent he had for years.

At the University I stopped athletics in my sophomore year to concentrate on academics. They were tough. I still, however, kept up my writing and radio work. In my senior year in M&P I was selected as one of three members of the University of Toronto team for the William Lionel Putman Mathematics Competition against major universities in North America. We beat Harvard and MIT.

I was awarded fellowships to Harvard and MIT. I turned them both down and went instead to the Institute of Aerospace Studies located within the realm of the University of Toronto. I now serve on the Advisory Board of the Institute.

I started grad school in 1951. My first effort at discovery was in the summer of 1952 when I completed the work for my Master's degree and was working at the Institute of Aerospace Studies calibrating a shock tube. It was a dark afternoon and I saw a glow in the tube. I examined the tube and saw that the particles of diaphragm material, which separated high and low pressures, and which was punctured to create the shock wave, were charred. I wondered at this. I did a bit more studying and examination and came to the further conclusion that the temperatures within the shock tube had achieved sufficiently high values to char the plastic diaphragm material. I varied the manner in which we were producing shock waves and deduced that we were ionizing shock waves when they reflected from the back of the shock tube. The temperatures exceeded 100,000 degrees Kelvin. On further study, I came to the conclusion that we were studying a process that was unknown at the time. This was a process of ionization particle acceleration. It could be used for space travel or for increasing the compression ratio in car engines. In fact, two of my classmates went to work in the Ford Motor Company to work on such a project. This has since become known as Plasma. The application to space travel is known as "Plasma Propulsion." (There is a wealth of information about this on the web.)

This has great promise in space, since no fuel is required. It is a major step forward. In addition, the concept of small incremental but continuous thrust can lead to spectacular velocities.

I became intrigued with this concept of plasma generation with shock waves and saw a great potential for space flight. The Associate Director of the Institute came to me and told me to consider this for my doctorate study. I thought it was a good idea until he started to load it on with enough work for 10 people for 10 years. I finally cried "uncle" and went to the Research Director, Dr. Gordon Patterson, and was assigned a new project. Sitting quietly in his office he told me that the project he had in mind was to determine the heating effect of space vehicles on reentry into Earth atmosphere. We had equations of motion we could rely on at sea level and in outer space. We had nothing in between the so-called "Slip Regime."

There was a technique advanced by NACA, the forerunner of NASA. This was called the Rayleigh Method. Rayleigh's method of dimensional analysis is a conceptual tool used in physics, chemistry, and engineering. This form of dimensional analysis expresses a functional relationship of some variables in the form of an exponential equation. It was named for Lord Rayleigh.

I accepted this project and soon determined how to solve the problem but needed significant calculation capability. Although this was the infancy of the computer era, we had one of the first computers in the world at the University of Toronto. Built by Ferranti International, an electrical engineering and equipment firm in England, it arrived in Toronto in 1951. I took courses in Computer Design as part of my Master's year, and learned how to program. We called our machine FERUT, for Ferranti at U of T.

I started using the computer at the University of Toronto, using some of 140 hours of machine time after midnight. The end result of this work was to determine that the Rayleigh's Method did not work. I generated a solution according to the technique proposed and came up with curves that went in opposite directions; impossible results. Finally, I went to Director Patterson, who told me it was worth the PhD because I had demolished a theory that was considered to be the solution. I told him it wasn't enough for me but would go back and try to find one. I did. I took an empirical approach and combined the equations at outer space, the Maxwell Equations, and the equations at sea level, the Navier-Stokes Equations, and linked them with parameters that would apply in the Slip Regime. My solution was a function associated with the Knudsen number, which was a ratio of the mean free path at a specific altitude over the dimension of the space craft. The mean free path is the average distance a molecule travels before colliding with another molecule or a space body. When I produced my results, it was a continuous curve and there were experimental results on the curve. I went back to my Research Director. He agreed that I had solved the problem. By determining the effect of the heating of space vehicles upon their return to Earth. I asked him if this was worth a second PhD. He laughed. This gives you an idea of the relationship I had with Gordon Patterson. He was a tremendous man with a wide-ranging sense of humor. He was also quite brilliant. I was fortunate to have such a good man as my Research Director.

What I had really done was to establish the parameters for the creation of heat shields which would make space flight possible. This was recognized in a report in the NACA by a man called Streeter, suggesting that the Martino Rule be used for calculating the heating effect of space vehicle returning to Earth. Patterson was also quite enthused and gave interviews of this result to newspapers in Toronto.

By the summer of 1954 I was fairly done the work required for my dissertation. Dr. Patterson decided that all PhD candidates should have at least three months as an intern somewhere. I went across the street to the DRML building – Defense Research Medical Laboratories – a division of the Defense Research Board of the Department of National Defense of Canada. I was welcomed with open arms by Dr. Les Turl, I think he was the Director of Arctic Warfare at that time. He directed me to examine a particular project which was top-secret. He assigned me an office that I would share with the Director of Chemical Warfare. Our work was top-secret, with Mounties at our doors for security. We also had daily burn bags for carbons and discarded worksheets of our project.

My project was associated with Arctic Warfare. I got so involved in this that I did not leave until November. I wrote my report which was labeled top secret and hence I could never see it again unless I was cleared for that project on a need to know basis. I've often wondered what happened to my work. As I left, my colleagues all wanted me to stay and hoped that I would apply to work with them when I was awarded my PhD. I told them the truth, that I hadn't been concerned with that aspect since I was concentrating on earning my PhD rather than a job afterwards. I thanked them for their consideration and for making me feel so welcome. I was looking for a position close to my aerospace training. They understood.

In June of 1953 a conference on Heat Transfer in Space Flight or some such engaging title was announced at the University of Michigan. I was given permission and financial support to attend. Upon arrival I was greeted by the director of the Aerospace Institute at the university and we shared experiences and results of my work area. That night I shared a room with a top ranked German scientist from Peenemünde. He filled me in on their efforts during the war and afterwards. He was rather famous but unfortunately, I cannot remember his name. He worked directly for Wernher von Braun, the father of the V-2 rocket for Germany.

The next day it was discovered that the research lab had burned down overnight. I was shaken because for safety reasons I had left the only copy that existed of my dissertation in the lab. I sort of scratched my head a bit. Those were the days when there were no copiers, the only copying system being photographic with chemical development of the end product. It was tedious. Hence copying was not a very common capability. I recall at the time that I did have detailed notes and could reconstruct my thesis; which I did upon my return to the Institute. It took me two weeks and taught me a very important lesson. From then on one way or another I made sure that I had a copy of whatever I was producing and that the copy was in a safe fire-proof location. Of course, today that is not a problem at all with everything computerized or very readily copied.

Despite the difficulty I experienced because of the loss of the only copy of my thesis, my trip was an outstanding success because of the information I gained from my discussions with the director of the University of Michigan lab, and with my many hours of discussion with the German scientist, who I recall was a Professor of Mathematics at the University of Berlin before the war.

After passing the Departmental and Senate Oral Examinations, I was eligible to receive my PhD. I was forewarned by the Director to be succinct during the Senate Oral and to keep my explanation to a maximum of 30 minutes. When I entered the room for the Senate Oral, I found that the Director had packed it with all the top faculty members – the Dean of Engineering, the Heads of the Mathematics,

Applied Mathematics, Physics, and Aeronautical Engineering Departments. I figured he was showing off a top student. I think they felt the same way. I was on the spot to perform. I was winging along and looked at my wrist watch and saw 29 minutes gone. I finished my sentence, put the chalk on the ledge, and sat down. I was about a third of the way through.

I thought Dr. Patterson would split a gut from inward suppressed laughter. Only he and I knew what I had done. All the others nodded wisely and complimented me for the brevity of my presentation. They commended me on my work. I wonder to this day if they had any idea of what I was talking about. I was way off in space, to say the least. It was "heady" stuff. But I had my PhD!

I joined a firm called Adalia in Toronto, headed by the man who invented radar, Sir Robert Watson-Watt. He gave me my first assignment, as the head of the Aerospace Division, a project associated with long range navigation or transatlantic flight. This was a system invented by the Decca Navigator, a division of the Decca Record Company. The system was called Dectra and was perfected by me to produce a quality position that could be used for transatlantic flight. I refined the technology even more, so I could determine the altitude as well as the position of the aircraft. I used Bessel functions and produced a report indicating how this could be done by taking into account the variation in the conductivity of the radio waves over the ground no matter if it was water or land.

Through Dr. Cally Gottlieb, the Director of Computer Center at the University of Toronto, I learned Remington Rand was opening a Univac division in Canada. I returned to Toronto and applied for a senior job in that new group with Jesse Greenleaf and his boss Ed White, who was the sales manager of Remington Rand. They would head this project. They hired me and I suddenly became involved in hiring a significant number of people and training them pending the delivery of Univac II system for my use as Director of the Data Center in Toronto.

The intractable thing was the difficulty in introducing new programs because of the computer language. I became involved with Dr. Grace Hopper, the high priestess in a sense of programming, who had developed the various forms of automatic programming. At the University of Toronto, I had used a system called Transcode and had developed my own system called EasyCode which simplified dramatically the speed in which programs could be written. My language was also quite close to the normal vernacular. Later in life, I produced something called JOE (Just Ordinary English) as a means of writing English statements which were then compiled or converted into computer code by the computer itself (the forerunner to modern-day search engines). In any event I worked with Grace Hopper with her English language program called FLOW-MATIC. We applied this in full detail to the installation of Univac II systems at the Ontario Hydro-Electric Commission. This was very successful, and we significantly reduced the cost of programming.

I reported this to various technical societies. In the meantime, I had become quite friendly with John Mauchly, the co-inventor of ENIAC. He told me that he was being virtually discharged by ENIAC and wasn't too sure what he was going to do. I convinced him that we should start-up a company and that I'd go with him. The company was called Mauchly Associates and we had as our product something called the Critical Path Method. John had headed up the Remington Rand Applied Research Center and as part of their effort they had developed something called Construction Scheduling. Morgan Walker of DuPont and Jim Kelley of Remington Rand had worked as a team. John and I hired the two of them to work with Mauchly Associates and very rapidly we took the construction technique and converted it into what we called the Critical Path Method. My development efforts in this technique was to create automatic ways of scheduling resources to flatten the requirement into a smooth curve line instead of an erratic curve. This would significantly reduce the cost of completing any project. I used this technique to secure contracts which very rapidly made the Critical Path Method something that became common place in use. This led to the attraction of the Olin Mathieson Corporation which had significant amounts of construction underway and they wanted to apply the Critical Path method to some $140 million under construction. A significant number in 1961. I was actually on my honeymoon in England when I got a call at 5am in London to offer me the job. The man calling me had lost track of the difference in time and which way it went. I didn't mind because it was a significant offer. I accepted.

When we returned from our honeymoon, I cleaned up my affairs in Toronto, as indicated in my memoirs, then we moved to New York. At Olin Mathieson I had my hand in various things that were invented. From a technical point of view the most important was the creation of a decision table translator, which was the technique for creating computer codes from logical decision tables. This was a very important technical development.

*   *   *

My usual contributions to inventiveness came from asking questions. When I visited the Squibb plant, I found they were making toothpaste by combining pumice and a flavoring material with milk of magnesia and hanging it in a bag to let the water drip off. I burst out laughing when I first saw this and suggested they take magnesium sulfate, add the pumice, add the flavoring, a little water, and stir. They did. That saved a lot of time and trouble in making toothpaste. Some other things I did that saved some money were such as inhibiting the need for a $120 million soaking furnace for aluminum ingots when we already had two. It turned out that they were filling these furnaces with the same type of ingots when this was unnecessary. It was a simple thing of asking the right question: "What happens if you overheat an ingot?" The answer was "Nothing." So, I told them to load the furnace with other ingots. This solved the soaking bottleneck and saved $120 Million.

I left Olin Mathieson after being recruited by Booz Allen and Hamilton to be the Director of their Special Projects Division. What they didn't tell me during the recruitment process was that I would be spending most of my time crossing the Atlantic. In particular, I would be spending more time in London than just about anywhere else. I was newly married with a one-year-old son and this soon became a very tenuous occupation, especially when I had to take my son to the hospital for a surgery, wait until he recovered, and then fly back to London to go back to work. While sitting in the hotel room that night, I decided to leave Booz Allen despite the attractiveness of the job and the monetary aspects. I resigned and returned to Philadelphia where I established my own consulting company, R.L. Martino & Company. The first year in business my total income was less than my income tax the year before, but at least I wasn't traveling to London like I was going downtown. I kept progressing, getting contracts and then a tremendous opportunity emerged. I was asked by Girard Bank if I could generate an electronic funds transfer system to replace their teletype system. I knew nothing about this area and soon learned that all the banks had teletype systems for transferring money. If you wanted to move money from one bank to another, you would send a wire or a teletype and that's the way it was done. It was heavily personnel bound and quite noisy. The wire transfer rooms were notorious for the sound level. Not only were the teletype machines going, but people were constantly barking on the telephone confirming what was desired. The way money was moved was that a telegram was sent which generated an electric tape that was ripped off one machine and put on another machine and sent to a bank. I studied the situation and determined that I could not only replace the entire teletype function electronically but could vastly increase the security. I designed a system using Wang computers and an electronic display screen into which the information for the wire transfer was entered. To a large extent I used repetitive information which was stored and need not be entered again, thereby minimizing the amount of data that was actually entered at any given time for a wire. There was a complexity at that time that we had a system called Bank Wire and another system for the Federal Reserve – so wires were either "bank wires" or "fed-wires." Thus, I created the first ever Electronic Funds Transfer (EFT) System for Girard Bank, now part of Mellon Bank East. Later in working with the Girard International Bank we came up with the need to do the same sort of thing with SWIFT, the international system for the exchange on money internationally. But that is a later story.

As I produced the system for Girard Bank, I interjected a simultaneous double save. The Wang system had that capability, so I used it. What that meant was that every transaction was stored twice on the computer for safety. I also interjected certain rules with regard to verifying the accuracy of the transaction. I had every wire transaction entered by two different people at two different locations on two different machines. Or I should say two different terminals, since all the information went to the same central computer. I also had a release process with a third person who would actually initiate the movement of the money. This created greater safety and security than had ever existed in the past. There was a great deal of discontent among the operators of the teletype system against such security measure since it resulted in more time being taken. We prevailed because our system actually reduced time and personnel with our system.

Later wire transfer systems would have people sitting in cages with their computer totally isolated from other computers. In addition, the central storage was duplicated among different machines in different locations so that there was a safety measure in terms of data protection against technical loss.

*   *   *

In 1964 or 1965 I was present at a seminar on Management Information Systems in New York – over a hundred senior executives were in attendance. One was Rear Admiral Basil N. Strean, Deputy Chief of Naval Personnel at that time. "Smoke" as I came to know him, later became a Vice Admiral in charge of all Naval Air.

His attendance at the Seminar led him to ask for a proposal to direct a Task Force created to design a Navy Manpower Information System, or NMIS, which exists to this day so far as I know. I am sure, however, that it has been substantially upgraded from the system we designed at that time.

The Task Force was composed of Senior Captains and Commanders. Many became Flag Officers. As a result, during the 1960's I spent four years in the Pentagon or in the Navy Annex working on projects for the U.S. Navy.

The highlight of this tour was a week we spent at Camp Ritchie, around the corner, so to speak, from Camp David. They didn't warn me about the "cannon." At 5:30pm a cannon went off and then the bugle sounded. When the cannon went off, I must have jumped a foot or more. Everybody had a smile on their faces when they asked how I liked the cannon.

The report was exceptional! It was one of the most interesting and most rewarding projects I had the privilege to lead. The "team" was tremendous. During that time, I also had the opportunity to spend a lot of time with Grace Hopper. She had an office on the fifth floor of the Pentagon, under the eaves so to speak. I regularly reported to SASN, Special Assistant to the Secretary of the Navy.

*   *   *

After the Navy effort I had other projects that I worked on; most especially one for American Express Credit Corporation. For them I developed the first online, real-time Commercial Paper Trading System. They were able to use it for strategy as well as operations. It gave them an enormous competitive edge in the marketplace. Word got around that they had a new system. They were not reluctant to give my name. When the Treasurer of IT&T, Gene

Girard, heard of this, he called me up and asked if I could install a system up and running for him by January. It was November. I said sure, thinking I had 14 months to complete the project. Then he told me he wanted it done in two months. We worked every day except Christmas day and had it running by January 4th.

Prudential also learned of our work and asked for a proposal, which led to a contract to build a system for them, with even wider capability.

The epitome of these projects was the system we created for General Motors Acceptance Corporation. Because of our success with Girard Bank, Prudential, and American Express where we created the first online real-time commercial paper control system, we were asked to be one of the seven bidders on a system for GMAC. All bidders were very large national companies. We won. My design was rather unique. It consisted of two centers, which we picked to be in Detroit and Dayton. The centers would have telephone lines coming in from two separate exchanges so that if one exchange went down, we still had continuity with the other. In each data center there were a number of computers. These were Tandem Computers. They were self-checking. They would also check for redundancy and continuity of operation if one machine went down. We actually put 12 Tandem Computers into each data center and created a system of duplex write on every transaction. When a transaction was consummated, the originating terminal was not released until the transaction had also been saved in the other data center, and hence there were duplex saves in duplex centers in real time. It took three years to build the system and to check everything since we were automating many things that were done by hand and had never been done by computer before. The total value of the transactions was $10 billion per day which was significant at the time. We started in 1982 and finished in 1985. Everything checked out and we went live.

What is significant was that on the third week after we went into operation in 1985, a backhoe tore out the telephone lines going into the Detroit center, knocking it out totally so that the entire country was operating only on the Dayton center. There wasn't a blip. Everything worked perfectly. Recall that we were doing $10 billion a day. In 1985 this was significant. The financial centers on the nation would have been in turmoil if there were no control measures designed to keep the system going.

It was a non-event. Nobody told me about it for weeks. When I found out I laughed and used that in my presentations on the system.

I formed a company called XRT to succeed everything I had been doing up until then. This company was formed in 1972 so that was the contractual company for the GMAC system. XRT became the dominant firm in systems with commercial paper, which also included electronic funds transfer. When we installed the system for Aramco many of the wire transfers exceeded $1billion. We installed security measures that to this day are still effectively in use. By the time I sold my interest in the company in 1997 we were moving $3trillion per day through our systems and not one penny had been diverted by anyone. We had actually involved ourselves in this together with Exxon. They hired a top-notch firm to try and penetrate our system. We watched everything they did and found that they were continuously on the wrong track. They didn't know they were working for us. But I found the study of interest in indicating that we had found a way that still, to this day in 2019, has not been penetrated.

It was 1994 when I became very interested in the idea of a Smartphone. This led to the filing of design patents and a utility patent that I filed on May 19th of 1995. This was the forerunner of all Smartphones. As a matter of fact, there are features in the patents that still have not been pursued on Smartphones. A companion book, *The CyberFone: The First Smartphone*, is being written by Joseph Martino and Joseph Looby. It will concern itself heavily with the design process and patents of the Smartphone.

For now, I will say that my track record of inventiveness was a forerunner of inventing the machine that changed the world. It created a trillion-dollar industry, and at one time I thought I would end up as one of the richest men in the world. By the time I was done with the major corporations that controlled everything, I was lucky to still have my shirt. Suffice to say that at this time that Frank Whittle, the inventor of the jet engine, lost control of his patents because he didn't have the money to pay the fees to retain the patents. Rudolf Diesel who invented the diesel engine, died penniless and bankrupt. Sir Robert Watson-Watt who invented radar had to proceed with an action in England's Privy Council, the equivalent of the United States Supreme Court, to find that he was indeed the inventor of radar. John Mauchly and John Presper Eckert who jointly invented ENIAC, the first digital electronic computer, received virtually nothing for their invention. In fact, there was a later action by Honeywell that did not want to pay royalties on the Eckert-Mauchly patent, then owned by Sperry Rand. The case became famous for the amount of "paper" it generated. Honeywell unearthed John Atanasoff who had built the ABC Computer at about the same time. Honeywell gave him $400,000 to build a modern version of that machine. It did not work. Despite this, Judge Earl Larsen ruled against Sperry and questioned whether Eckert and Mauchly had truly been the inventors of the digital computer. The

Federal judge, Earl Larsen, was totally ignorant of any of these matters. He was confused about the difference between digital computers, analog computers, and machines that could operate one program or a number of programs. ENIAC was a general-purpose digital computer. ABC was a single purpose machine. Larsen sided with Honeywell, and selected John Atanasoff, someone else to be given credit for the invention. The Eckert-Mauchly patent was invalidated. The full details, and my considered opinions, are included in my book, *People, Machines and Politics of the Cyber Age Creation,* published by BlueNose Press in 2007 and distributed by Amazon.

My history of patent suits had a different outcome. In my case I sued 161 companies: 101 settled, and the others went to the White House. The interpretation of the law was changed so that I, as the inventor, could not exercise the right of my patent because I was not a practitioner. I didn't have enough money to be a telephone company and that did it. The "big" companies won. Suffice to say, I claim to be the inventor of the CyberFone, the first Smartphone. The prototypes I built from 1995-2000 were the first Smartphones anywhere. One of them won the Best of Show Award at the Sun Microsystem Java One show in San Francisco, June 15-19, 1999. It was during the time of the show that

I travelled to Cupertino and met with a man who I think was Tim Cook.

CHAPTER 3: FRIENDSHIPS

During my lifetime I have met many interesting people. Some were in high places in the church, government, or the military, and some were what might just be called ordinary people. The military officers included Air Vice Marshall Fred Johns of the Royal Canadian Air Force; Colonel Edward Churchill, Canadian Army, Director of Special Projects; Vice Admiral Basil Strean, Chief of Naval Air; Admiral Edward Watkins, Chief of Naval Operations United States Navy; General P.X. Kelly, former commandant of the US Marine Corp and a member of the Joint Chiefs of Staff, plus hundreds of others in military service.

Among the famous were the three Popes, St. John Paul II, Benedict XVI, and Francis I. Barbara and I met each of them numerous times. In particular we met Pope St. John-Paul II at least 30 times. He came to know us and repeatedly asked us to give our regards to Cardinal Krol, his good friend, and to our children. We were privileged to attend Mass four times in the Private Chapel in his apartment, and on one of those occasions Barbara did the readings. He was startled to hear a female voice. After Mass, his secretary Monsignor Stanislaw "Stash" Dziwisz, who later became Cardinal Archbishop of Krakow, patted her on the shoulder and said, "You done güt!"

After one private audience with Pope St. John Paul II, I introduced our sons Peter and Paul to him, and then went around and introduced them to Dziwisz. He was absolutely thrilled. He had become a close friend over the years because of our relationship with the Pope on other matters. I remember one time I breezed into Rome and phoned him the night before to see if I could attend the Pope's Mass the next day. His voice, full of regret, oozed with apologies. The next morning at 6am he phoned full of enthusiasm, and said, in Italian, "It is possible. Arrive at the bronze door by 7am and you know where to proceed from there." The bronze door is the entrance from St. Peter's Square into the papal apartments and is guarded by Swiss guards. From there you proceed to the office which gives you entry into the papal apartments and you climb the stairs until you arrive at the Pope's apartment. There is an elevator from the papal courtyard as well.

The Pope's apartments are spacious. There is a very large library where he works and where he holds many of his private audiences. Barb and I were privileged to be in that room with him on many occasions. We were even highly privileged to have a private tour of the apartment by Cardinal Harvey, head of the Papal household. Off this room is the entrance to the private chapel which holds about 40 people. It is usually crowded with the papal staff and any other prelates who happened to be visiting. In the early days of the Papal Foundation, the Stewards would visit the Pope. A select number of the Stewards of St. Peter, members of the Papal Foundation, would be permitted entrance into the Mass by the Pope. To say the least, attendance at that Mass is awe-inspiring.

I have met many rich and powerful and famous people. These have included President George H. W. Bush, whom I met in the Oval Office, and Ambassador Walter Annenberg, who had been the Ambassador to the Court of St. James, Great Britain, during the presidency of Richard Nixon. That is where he met his wife Lenore who at that time was Chief of Protocol.

The largest group of people I knew were priests on one hand, and on the other inventors and creators of the computer revolution. The inventors included Sir Robert Watson-Watt, the inventor of radar, Dr. John Mauchly co-inventor of the ENIAC the first electronic computer, Dr. John Presper Eckert, co-inventor with John Mauchly of ENIAC; and Dr. Grace Hopper, creator of many of the systems for creating computer code using the vernacular and using the computer itself do that translation. Grace was a serving officer of the United States Navy when she went to work for John Mauchly on the ENIAC project, and then went back to the Navy where she ultimately rose to the rank of

Rear Admiral. A building in her name is located the campus of the U.S. Naval Academy. By the way, I was also very friendly with Kathleen "Kay" McNulty Mauchly, John's wife. She had been one of the six women hired for the ENIAC project. As such, she was one of the first programmers ever in history. She was also the person responsible for me meeting Barbara. I recounted in Chapter one the story of how she asked why I wasn't married, and I said John kept me too busy and she said, "I'll fix that." And she did.

My favorite people were my priest friends. How can I even mention them all? I have included significant information about many of them in my autobiography. For now, however, let me tell you about Delmar "Del" Skillingstad, SJ. He was certainly an alter-ego. He had spent four years as Treasurer of Sogang University in Seoul, South Korea, followed by 16 years as President. I came to know him rather well when he was President of the Gregorian Foundation. For years, he and I monitored and directed the activities of the foundation portfolio. We were proud to report each year to the Board that we were increasing the portfolio by between 20-30%. We took the initial $3million with some incidental contributions along the way and raised it to over $100million.

Del was another close "brother." There are so many good times that it's hard to recall the many that were outstanding. I remember his 50th Anniversary of Ordination. He and I were Scotch connoisseurs or so we thought. Anyway, I went out of my way to get 50 small bottles of Scotch, all of them different. Try it sometime, it isn't as easy as you might think. At that time, I was traveling all over the world and picked up quite a few bottles on the flights. The final bottles I picked up in a store in Scotland.

Del would often visit us, especially every summer to spend a week or more with us at the shore. During this time, we always went to Atlantic City and had our evening in the casinos. We would play blackjack until we won enough money to buy dinner. Notice I said "won." We always did. We are great blackjack players. We also had some grand and glorious dinners with associated aperitifs, usually scotch. In between, we discussed the financial markets. We had developed a technique of picking areas and industries rather than specific stocks. We had as our advisor Chuck Clough, who at that time was the Chief Investment Strategist for Merrill Lynch. He now runs the Clough Hedge Fund in Boston. Chuck is very successful. He lived in Boston and worked in New York during our association, commuting between Boston and New York. All of our meetings were in New York.

Del came to the weddings and baptisms of our children and grandchildren. An interesting wedding that he presided over was that of our son Peter, a graduate of the U.S. Naval Academy. Peter wanted to be married in the West Point Chapel because the entire wedding party was staying in Arden House, a property close to the chapel. His prospective father-in-law was on the board at Columbia University and Arden House, where the wedding party was staying, was owned by Columbia. Dwight Eisenhower, at the time President of Columbia University, had stayed there, and Peter stayed in the room that had been occupied by General Eisenhower.

When Del left New York, he was President of the Gregorian University Foundation. He moved to the West coast and became Assistant to the President of Gonzaga University. At that time the President was Robert Spitzer, SJ. Del introduced me to Bob and told him about the Smartphone, which I called the CyberFone. Bob became intrigued. He immediately saw the scope of the benefits and impact of this. I could almost hear the wheels turning when I was talking to him. The end result is that he went to bat and had me awarded an honorary degree by Gonzaga University. I flew out to Spokane to receive this. Immediately afterwards, the same night, I set out on my flight to Vienna. There was an important meeting I had to attend there and then from there I flew to Moscow.

Father Skillingstad and I were close in age. We often talked of death and our hopes for more years to finish what we were doing. Del died very quietly one day. His Superior told us he was ill and just went to bed. He died two days later.

I have some fond reminisces of my times with Cardinal Keeler. We often had lunch or dinner when he was in Rome. Fairly often we were joined by Archbishop, later Cardinal John Foley. We always had a great time. When Keeler was not in town, we would have dinner with John Foley. Often, we would end the evening with a trip to his famous apartment in Villa Stritch. We also had dinner often with Keeler in his residence in Baltimore.

I had the occasion to meet Cardinal Keeler quite frequently when he formed the International Committee for the Restoration of the Basilica in Baltimore. He asked me to be a member of this and then later asked me to head it. On more than one occasion I suggested to him that he come sailing in Annapolis and he agreed to do so. He came one Friday evening and we went out in my sailboat. After things settled down, he said, "I want to talk to you. Let's go down to the salon." We went down and he proceeded to ask me to head the entire program for the restoration of the Basilica in Baltimore. I was awed and overwhelmed because of the significance of the Basilica. It was the first cathedral in North America. It was the one that had the largest dome designed by Benjamin Latrobe who later designed the dome for the U.S. Capitol. The Basilica building was certainly an historic edifice. The objective was to restore it to its original grandeur without all the artifacts that had been added to the building since it was built. The building had been dedicated in 1806, and the objective was to have the new building, or the recast building, dedicated in 2006. I agreed to do it.

Our objective was to raise $35 million for this project and I became heavily involved in this fundraising, and succeeded, introducing Cardinal Keeler to potential donors.

I also worked with the architects who were quite famous. They were the ones who had restored Grand Central Station in New York and were experts at restoration. But there was a problem with the Basilica. Everyone wanted to reactivate the church's basement, but the ceilings were quite low. I looked at the plans and suggested that walk-ways could be designed to come between all the foundation pillars. This gave us an extra foot and a half making it feasible. That way we had walkways that could proceed to various seating areas which were still subject to the six-foot ceiling. As we proceeded, I made other suggestions of that nature with my engineering background and had a tremendous working relationship with the architects.

Working with Keeler was a joy. However, no matter how often I came to the residences, he'd ask "Did you sign the book?" And he insisted that all guests at any time no matter how many times, sign the book. I was pleased to do so. At receptions it was also my pleasure to get his favorite drink, Compari and Orange Juice, from the bartender.

Cardinal Keeler and I spent many an evening discussing world affairs, and with whom I established a relationship almost that of a brother. I stayed in his home and he stayed in mine. He attended our 50th wedding anniversary and my 80th birthday.

I often sat quietly with him over dinner in the residence. Just the two of us. We would listen to music, discuss the art in the building, and any other matter that was current. We were two buddies. I often stayed in the residence. After a while I wondered if I had a fixed room there, since I always got the corner bedroom in the Northwest part of the building. It was comfortable and quiet.

After the building was dedicated, our friendship continued. He came to Palm Beach to stay with us and came to my lecture with George Weigel in Philadelphia.

My son John and wife Susann had twins born prematurely at the 27-week gestation, and when I visited the day after birth, I saw the dire condition of their health. I phoned Keeler and asked if he could arrange to have immediate conditional baptism in the hospital. Later that day, John was walking in the corridor when he noticed someone dressed in clerical garb at the telephone. The man hung it up quickly and virtually ran from the phone. It turned out that he was the Catholic Chaplin; who then proceeded to administer conditional baptism to the twins. Keeler moved quickly.

After the twins' sister Ann was born, we had a ceremony in the Basilica with Cardinal Keeler who, in full regalia, baptized Ann and completed the sacrament for the twins, Thomas and James.

It broke my heart when I visited Cardinal Keeler at the home operated by the Little Sisters of the Poor. He couldn't communicate at that time. But when the music started, he could sing the ditties. This is something the nurses and nuns told me that through music they could communicate with their patients. He died shortly after our visit. I attended the funeral and felt that I was losing a brother.

I was very friendly with other Cardinals as well. But nothing like the closeness I had with William Keeler. Probably the one closest other than him was John Foley. And so, I could go on and on. I've been blessed with many friends over the years. They're all gone. I miss them terribly. I think my survival from many difficult situations, including terminal cancer, is due to their prayers and support. I cannot however forget that throughout my life I have always felt that my mother, in particular, was looking after me. I could swear that she stood at the corner of my desk feeding me answers to the questions on exams that I was struggling with.

Michael Joseph Smith, SJ was another great friend. The most important thing I did following his death was in a Board Meeting of Saint Joseph University. One of the items on the agenda for that meeting was the naming of the new chapel. There was a significant move by some members of the Board to associate the naming of the of the Chapel to raising $5million. I, on the other hand, was totally determined to name it the "Michael J. Smith Memorial." It was a discussion that went on all day long. I won the support of the board to the extent that one of them finally said to me "Okay Rocky, make your motion." I did. I am grateful to have made the motion that made the Saint Joseph University Memorial Chapel as the Michael J. Smith Memorial Chapel at St. Joseph's University.

Jim Maguire's wife Frannie sculpted a bust of Mike. This is at the back of the chapel. It doesn't exactly look like Mike, but it is enough of a resemblance. For those who didn't know Mike it doesn't matter because the bust is quite impressive.

Father John "Jake" Laboon, SJ was my spiritual advisor for a few years. He was a Navy hero, decorated during the Second World War, in which he served as Executive Officer on a submarine in the Pacific. After the war, he was designated to receive his own command, but he resigned to become a Jesuit. After ordination, he went back to the Navy as a Chaplain. He rose to the rank of Captain and was selected for Flag. John O'Connor was a Flag Officer Chaplain at the time. There could not be two Catholic Admirals heading the Chaplain Corps at the same time. John offered to resign but Jake would not let him. When Cardinal O'Connor recounted the story at Jake's funeral, he suggested that Jake was the better man. Jake has a Frigate named after him. When he died, the Navy wanted to bury him in the Naval Academy Crypt along with John Paul Jones. Jake turned it down, wanting only to be buried with his brother Jesuits. Jake was always my hero and Peter's. He commissioned Peter upon his graduation from the Naval Academy. The tradition is that any serving or retired Naval Officer can commission a Naval Academy graduate.

I have so many good memories of so many good friends. Arturo Lozano, SJ the Mexican Jesuit, was also so full of life whenever we were together. Archbishop Joseph Pittau, SJ – what an intellect, what a mystic! Cardinal Edward Egan, a buddy over many years; Cardinal John Foley, perennial dinner companion in Rome and later in New York and Philadelphia. Cardinal Krol of Philadelphia; Cardinal O'Connor of New York; John Blewett, SJ, mystic; John Dressman, SJ, military to the core; George Aschenbrenner, SJ, philosopher and theologian; and Domenic Maruca, SJ, moral theologian.

How can I ever forget John Snyder, SJ? John came to see me when he was 92 because he wanted to see me before he died. When he returned to St. Louis, he would undergo complex heart surgery. He survived and phoned me with an ebullient message telling me so. His last phone call was on his 96th birthday. I missed his call because I was out. I saved it and listened to it for many months thereafter. John died the day after. I phoned to tell him how much I enjoyed his message and to congratulate him on his birthday. He was dying at the other end of the phone as reported to me by the nurse-nun in charge. He died quietly during the call. I will never forget John.

Then there is Edward "Eddy" Nowlen, SJ. He had a PhD in Theology from Harvard and was a Professor at the Gregorian University. We walked all over Rome on more than one occasion, eating ice cream cones, and commenting on the passing scene. Ed had been a saxophone player in an orchestra; so, had John Snyder, played clarinet. Both played on cruise ships during their novitiate years. To say the least, they were ebullient full-bodied people.

Nicholas Rashford, SJ was named President of Saint Joseph's University in 1986. Mike Morris was Chairman of the Board. Nick and Mike asked me to join the board the following year. I spent nine years on that board, and hopefully I had some impact on the formation of the modern version of Saint Joseph's University, which is very successful today. It's not that the old university wasn't successful, but it was sort of static and certainly of the highest quality, but not really directed to becoming a leading university in the current environment. As matter of fact, the business school, the Ervian Karl Haub School of Business, which was an add-on at the time, is now quite famous in its own right. Nick did all of that. He put drive and *umpf* into the image of Saint Joe's. We became close friends. We often would have dinner on a Friday night when Nick and I would commiserate with our inability to eat seafood because of our allergies.

One lasting friendship is that with Nick and Mike. We get together occasionally, and always have a roaring time.

Nick is train devotee. On more than one occasion we chased trains around Pennsylvania and took pictures of these trains at specific crossings. Nick rode the Transcontinental Canadian National Railroad from Calgary to the West Coast, over the Rocky Mountains. They even let him toot the horn and drive the train for a few moments.

At the time, Nick was also Chairman of the Delaware River Port Authority. I talked Nick into joining the board at XRT where he was valuable in terms of continually stressing the need to focus and the need to generate more money than we spent. He was also instrumental in me coming to know Brian Kelly who served among other things as Chairman of the Board at Saint Joseph's University, succeeding Mike Morris. Brian was also Executive Vice President of Verizon and he was instrumental in me getting the first prototype of the CyberFone built by CTDI and helped me meet potential users of my patents. CTDI stands for Communication Test Design, Inc., a major firm in that environment.

Alfred Jolson, SJ was the Assistant Dean of the Business School at Saint Joseph's University. He had an interesting background as a Jesuit, having served in Baghdad and in other parts of the world before coming to Saint Joe's. He had a Harvard PhD. Al loved a good breakfast and would often come over to our house for it. I asked him to join the board of XRT, which he did. He provided much influence once again in terms of focus and actually in the recruitment of senior staff, former students of his who were outstanding.

One day Al phoned at 7:30 in the morning and told Barbara that he was a Bishop. She thought he was kidding, but it was true. He had been appointed Bishop of Iceland. Apparently, his father was Icelandic. Al had been offered the job some 10 years previously but had turned it down. This time Cardinal Pio Laghi, who at that time was the Vatican Nuncio to the United States, said very quietly, "You are the Bishop. We've already checked with your superiors. They have concurred."

I remember one time he came to a meeting in the offices in the Sky Tower at Grand Central Station. During the meeting the fire alarm went off and smoke came pouring up. We decided to get down from the 65th floor. We got into the elevator with smoke coming up all around us. I made the comment that we were perfectly safe since we had the Bishop on board. Al began laughing raucously. We proceeded to the ground and as the elevator doors opened, we saw smoke all over and firemen in full regalia looking at us and wondering why we had ridden in an elevator which was expressly frowned upon. Laughing a little because of my comment about having a Bishop on board, which we didn't mention, we left. Al and I often talked about that incident.

A few years later, when Al was in Pittsburgh doing confirmations for Bishop Donald Wuerl, he had a serious heart attack. Bishop Wuerl told me he wasn't expected to live. I phoned Al and talked to him. He told me things looked really bad. As a matter of fact, he was planning the liturgy for his funeral. I told him that we would pray for his survival. He said it was unlikely, but he was gratified. He gave me his blessing for all the years that we knew each other. He died on the operating table two days later. We had a memorial service for him in Philadelphia and then he was buried in Reykjavik, Iceland.

There is an eerie story about his funeral. The Cathedral in Reykjavik had been equipped with an electronic organ. The chimes began to ring at his funeral and rang for three days. No one could determine why. There was no short circuit. There was no way they could turn it off. It just went on and on. The story was reported in the Reykjavik news and a copy was sent to me. I knew why the bells rang. It was Al.

One final story about Al. We had a New Year's party at home one year. I think it was a few days after New Year, Jim Quigley answered the phone when it rang. Al said, "This is Al Jolson. Is Rocky there." Jim quick on the draw asked if he was going to come over and do a few songs for us. When I picked up the phone, I told Jim that this was Bishop Al Jolson which startled Jim a little bit and he whispered his apologies. I started to laugh, and I picked up the phone and said, "Well Al, are you going to come over and sing a few songs for us?" Al was laughing his head off at the other end. Such was the relationship with Bishop Al Jolson.

Among the people that I became close friends with was Father Peter-Hans Kolvenbach, SJ, General of the Society of Jesus (the Jesuits). I would have never thought when I was growing up that someday I would know the General of the Jesuit Order on a personal basis. I did. As a matter of fact, I remember I was walking one day in Rome with my son Paul, when we met Father Kolvenbach. He stopped and we talked for some time, and he obviously knew me with numerous personal tidbits. As we went on Paul asked me who that was, and I told him. He was floored.

I corresponded on a personal basis with Father Kolvenbach, SJ after he retired as General of the Jesuits. I was saddened to receive a note one day that he had passed. I still miss his smile, his sense of humor, and his quiet sanctity and brilliance. There was one time I met him at the Rome airport. He was traveling extensively. I saw he was holding a coach ticket. I had a seat in business class. I offered to trade seats with him, but he turned me down with a big smile. He said no he didn't need more space than he had, and it was good for his soul not to have a luxurious seat. That was Father Kolvenbach.

I could go one and on. And so, I think that I was blessed with many friends over many years. For this I thank God over and over.

CHAPTER 4: MY INVENTIONS, CAREER AND PERSONAL LIFE

I have led a very interesting life. It falls into three categories. First and foremost, of course, is my professional career, and the inventions associated with it. My separate life is that associated with the Catholic Church. This was rewarding in many ways and I will fill you in on this. My third life is very personal. This is associated with the things I did when I wasn't working or when I wasn't doing anything with the Catholic Church. In many ways that is the most interesting part of my life.

I will restrict this chapter mainly to my inventions because that is the direction of all the parts of the book.

Why did I invent things? Why did I discover things? What motivated me?

It was my curiosity. I also "saw" things. I could visualize an entire system and manipulate the pieces in my mind. Then I could describe it step-by-step as I assembled or disassembled the system. The language of description could be English, differential equations, or system components. For example, for the State of Illinois, I perceived the immediate need for an integrated database. This led to the concept of a Relational Database System, and a Database Management System with the elements linked in a relationship pattern. For me, it solved the immediate problem at hand, a system for the State of Illinois.

What did I do? I invented. My inventions, as with Illinois, were usually byproducts of solutions to problems I was addressing. Hence, I am a problem solver.

My first step was always to define the problem and then to find the solution. The vital element here is to define the right problem. Often by defining the problem, I could "see" the solution.

I have been told that Einstein often said he developed the Theory of Relativity by imagining that he was a light beam travelling through space. His description of what he saw became the Theory of General Relativity. His language of description was the equations he generated.

Mozart is said to have claimed that he heard the music in his head and that he just wrote down what he was hearing.

My inventions began in graduate school, at the University of Toronto's Institute for Aerospace Studies. I developed the equations and procedures of the heating characteristics of rockets on their reentry into space. This led to the creation of heat shields which were used to ensure the passengers in a space vehicle would survive. In two instances where the heat shield failed, the capsules were lost and so were the astronauts.

How did I come to that solution? I had been assigned the Rayleigh's Method as a means of determining the heating characteristics of a body reentering the atmosphere from space. As I recounted this didn't work. I then proceeded to invent a solution. I developed a formula that had to be true:

S(x,y,z,t) * (Navier Stoikes Equatioins) + R(x,y,z,t) * Maxwell Equations = 0

All I had to do was find the functions S and R. After a lot of tinkering, and following a suggestion of a classmate to involve the Knudsen number, I came up with:

S = 1 / (1 + KN)
R = KN / (1 + KN)
Where KN is the Knudsen Number for that altitude.

I didn't see it at the time, but I did about 10 years later. I had taken the first step in finding a general solution to the Unified Field Theory that had plagued Einstein in the later years of his life. He never did find the Unified Field Theory he sought. Using the same approach, I used for my doctorate, I would like to propose that the Unified Field Theory can be shown to be:

$F_1(x,y,z,t)$ * (Equations for Electro-Magnetism)
$+ F_2(x,y,z,t)$ * (Equations for Gravity)
$+ F_3(x,y,z,t)$ * (Equations for Light)
$+ F_4(x,y,z,t)$ * (Equations for Relativity & Motion)
$+ F_5(x,y,z,t)$ * (Equations for Radiation)
$+ F_6(x,y,z,t)$ * (Equations for All Other Forces)
$= 0$

This should solve the Unified Field Theory search. In addition, solving these equations for Black Holes should lead to the Speed of Light as a variable, heavily dependent on the density of the medium.

At the Institute, I also discovered something which is now labeled as plasma. This is the ionization effect due to shock waves or electronic discharges within a vacuum. It is my belief in something that I have contended for some time – plasma propulsion is the answer for space travel. The important thing with plasma propulsion is that no fuel needs to be carried. Just think of that: **No fuel needs to be carried.** Considering the fact that we have the Hubble Telescope unable to maneuver because its fuel is exhausted, the importance of having no need for fuel cannot be understated.

One other aspect of plasma propulsion and a very important one, is that acceleration is continuous. As a result, enormous speeds can be achieved over time. This will dramatically shorten the travel time between points in space. The trip to Mars could probably be reduced to 10-14 days. A trip to the moon could be accomplished in 1-2 days.

During my high school years, I frequently attended Saturday evening lectures at the University of Toronto. These were conducted by the Royal Canadian Institute. One Saturday in the Fall of 1944, I witnessed the awarding of an Honorary Degree to Sir Robert Watson-Watt, without any reason given, although the men in robes seemed to know why.

After the war ended, radar was announced. The fact of its existence had been Top Secret through the war years.

Immediately after finishing my doctorate work, in the fall of 1955, I attended a cocktail party where the guest of honor was Sir Robert.

We hit it off immediately. On the spot he offered me a job as Director of the Aerospace Division of Adalia, his consulting company. I took it.

One of the first assignments was to direct the contract with Decca Navigator. As part of this contract, I developed techniques for not only using the Dectra system for transatlantic travel, but also established the use of these radio beams to determine altitude with some precision.

One thing I should mention is that I developed techniques for calculating complex variables and complex integrals during my work with the heating effect in space travel. This led to significant improvements in various calculation procedures, especially the use of orthogonal polynomials for the evaluations of the Faltung Integrals. These resulted from the application of the LaPlace Transforms or Fourier Transforms to complex differential equations in order to reduce the magnitude of the problem by at least one dimension.

In my work at the institute, I spent five months at the Defense Research Medical Laboratories. The Labs also housed the Institute of Aviation Medicine. There I was instrumental in creating specialized techniques to be used for Arctic Warfare. This was top-secret and probably still is.

In my work effort under the broad scope of projects competed by R.L. Martino & Company, a consulting company that I formed in 1965, I created systems for electronic funds transfers that were the first of their kind. These techniques have led to the creation of systems that did make it possible, and will extend that possibility, in international digital commerce.

With regard to the contract with General Motor Acceptance Corporation, I created security systems that have become the standard even for today. These security systems also included the work I did with creation of electronic funds transfer. The first of these systems was for Girard Bank in Philadelphia, which is now a part of Mellon bank of Pittsburg.

In the 70s, with XRT, I began considering ways that systems could be created without any programming. This led to the creation of APG, Application Program Generator. With this method it was possible for me to create an entire system without having to program it.

At this time, I also created something called JOE, Just Ordinary English, which was a report generator that operated entirely in the English language and is the forerunner to today's search engines.

The capstone in this inventive streak was the Smartphone. I called it the CyberFone. I also wrote articles about the Smart Society. I just wasn't smart enough to call my CyberFone the Smartphone. It took someone else to do that. But my CyberFone was the first Smartphone. I filed for the patent on May 19th of 1995. I built the first prototype in 1996. By 1999 I had the handheld device which is in common use today.

I was prevented from receiving my monetary due on the Smartphone by modification of the patent law. I tried in many ways to get legal representation for a lawsuit against specific infringers. In all cases I was turned down because the law firm I went to, had a client that was an infringer and would be affected by the suit. I had no choice but to approach patent accumulators, or trolls as they were labeled. This gave me the ability to have top notch legal representation in launching lawsuits against infringers.

We instituted action against 169 companies. 101 settled, and this gave us an $18million war chest in order to pursue major actions against major infringers. Some 68 companies refused to settle, went to the White House and complained that my actions were interfering with their ability to innovate and invent. Good grief, I was the inventor. I was the wronged party. But the greed of these infringers was beyond measure. They convinced the Obama administration to change the law and interpretation of patent law. The end result was that any organization that was not a practitioner would be denied the right to pursue its patent rights. I had been. But I ran out of money on one hand and was given some misrepresentation from a major chip manufacturer, with regard to the chips that were selected to transfer data from the telephone to the computer. In any event, I was no longer a telephone company and could not pursue my patent right. Some 69 cases were thrown out of court. As a result, I was essentially put out of business.

It is grossly unfair that an inventor should be denied by legal manipulation of some kind by the infringers. However, this has been a factor in history for many years. Frank Whittle the inventor of the jet engine, lost his patent rights when he did not have the money to pay the renewal fees. Rudolf Diesel, whose diesel engine changed the world of transportation died penniless and bankrupt. Sir Robert Watson-Watt who invented radar had to proceed with an action in the Privy Council in England, the equivalent to the U.S. Supreme Court, in pursuing his rights with the inventor of radar. Dr. John Mauchly and Dr. John Presper Eckert, who invented the first computer, were denied the fruits of their invention with a malicious lawsuit launched by an infringer, Honeywell, who refused to pay the royalties agreed to. Outrageous!

I have written about this in one of my books, *People, Machines and Politics of the Cyber Age Creation*, published by BlueNose Press in 2007 and distributed by Amazon. This is a detailed history heavily from my personal perspective, of the birth of the computer age, it includes a chapter on this malicious trial.

Now I will briefly tell you about my career – there is a brief biographical summary which includes a list of the jobs I held, in Appendix VIII of this book.

In April of 1970, I resigned as Chairman/ CEO of Information Industries. This was a company that I had organized after I left Booz Allen and Hamilton in 1965. We had been very successful in getting contracts that were profitable. The late 1960s was a period of excessive IPO activity. I was offered contracts for an IPO by various companies, most interestingly, by Allen and Company in. New York. I was also offered a contract by Robinson and Company in Philadelphia. I made a terrible mistake and accepted the Robinson offer. We went public in January of 1969 at $11 a share and raised $3million. It soared almost immediately to $20. As outlined in our prospectus, the funds from the offering were used to expand the company. We were set to become a major force in the growing Information Technology industry. Then we were hit by a recession that impacted us dramatically. The stock price dropped, and we turned from profit to loss. The underwriter with representation on the Board blamed me for the downturn. I was counseled to resign and did. The company was merged with a Dental Supply Company they controlled. Then it faded into oblivion. They had no concept of what I had done to set the stage for explosive growth. That episode soured me on public companies and underwriters.

I wrote letters to many people I knew. Almost immediately I was asked to write a proposal by the Department of Finance of the state of Illinois. It was accepted and I began in 18 months commute to Springfield Illinois. The result of that study, published as "Impact 70s," presented a plan for computer use throughout the state of Illinois that would result in cost avoidance of over $100million. As I stated already, this also led to the invention of a Database Management System.

This caught the attention of the Chairman of the Illinois Board of Higher Education. They requested a proposal which was accepted, but they wanted a corporate name, overnight I came up with the name XRT. I remember walking the streets of Chicago with my good friend John Gentile who at that time was Director of Finance for the State of Illinois. He liked the name immediately. I created the name because it sounded strong. X is the strongest letter in the alphabet, and I wanted to show strength.

I became Chairman of a Task Force that included the Presidents of the major public universities in Illinois. I met with most of them individually as well as visiting their campuses. I became very friendly with Dr. John Edward

Corbally Jr. who was President of the University of Illinois at that time, and later President of the John D. and Catherine T. MacArthur Foundation. Jack, as he was called, recommended me for the presidency of a number of universities, one of which I pursued. I almost made it.

My report to the Board of Higher Education was enthusiastically received.

This work led to a request for a proposal from the Federal Trade Commission. That report was accepted and implemented.

XRT was on its way.

Meanwhile, I became heavily involved in functions of the Catholic Church. Let me say that this falls under a limited number of categories. For the Equestrian Order of the Holy Sepulchre, I am now a Knight Grand Cross, and the recipient of the award of the Gold Palm, the highest award in this Order. The objective of this order is to support Christianity in the Middle East. My efforts were directed towards developing memberships in the Philadelphia region where no members existed at the time. This began in 1986 and the result was to build a cadre of about 200 members.

In 1988 I attended a meeting which led to the creation of The Papal Foundation. This meeting was conducted in the basement of the Papal Nunciature in Washington. I sat beside then Archbishop later Cardinal, James Hickey of Washington. Somewhere along the way he asked me if I was a Knight of Malta. I said no because no one had asked me. He smiled and just said "I just did." He then secured the support of Admiral James Watkins, at that time Chief of Naval Operations. So, both Jim's became my seconders for membership in the Order of Malta. I became a Knight of Malta in 1988 and did many things in the Order. I have been awarded the Cross of the Order pro Merito Melitensi, and I have also become a Knight of Obedience, a Knight of the Second Class. I have found my work with the Knights of Malta to be rewarding in many ways.

I became associated with the Gregorian University in Rome in 1981 when I attended a presentation on their limited use of computers. They had one PC. So, I asked why there wasn't more use of computers. By 1981 computers were quite common place. The answer was typically Jesuitical. I remember Father John Blewett saying to me, "Because you haven't come to show us how." I proceeded to spend a significant amount of my time going back and forth across the Atlantic, three or four times a year, computerizing the Biblical Institute of the Gregorian University. In this I was helped immeasurable by Del Skillingstad who carried a disk in his jacket pocket and would go around to the various computers that we had working for us and gave them a "shot" of DOS. This was rather humorous as we explained what we were doing. I became a member of the Board of the Gregorian Foundation. Within a short period of time I was the Vice Chairman. Peter Mullen, Esq., the managing partner of Skadden Arps in New York, the largest law firm in the United States, was chairman. Peter and I formed a relationship that went on for 16 years until I was forced to resign when I contracted lymphoma.

In Chapter 3, I've described my friendship with Cardinal William Keeler, Archbishop of Baltimore, and described his request of me to share in the restoration of the Basilica in Baltimore. This is probably one of the noblest things I've done.

I was asked by Cardinal Krol to attend the inauguration meeting for the Papal Foundation at the Papal Nunciature in Washington in 1988. I am a member of the Papal Foundation and as a part of this I have on numerous occasions met in personal audience the reigning Pope at the time. In addition, I was privileged to attended Mass by Pope St. John-Paul II in his private residence. This was done four times. I also met with him some 30 or more times. I have also met with Pope Benedict and Pope Francis.

The work I did for the Catholic Church took a significant amount of time. It was all very worthwhile. I not only had a feeling of accomplishment, but I had a great feeling that everything I did was truly appreciated, and something to be useful in human affairs. We spend so much time in commercial and political involvement that we tend to overlook the fact that fundamentally we're people, and it's very important for us to learn how to help each other and get along with each other. This belief came to me as a result of my work with the Catholic Church.

And then there is Rocky, the person. What did I do? What made me tick as a person? Well, I am proudest of the fact that I spent significant time with my sons as they were growing up and with my grandchildren now. Fathering is a very important element that is often ignored. Once again, I put significant time into this and received benefits that cannot be measured in any way at all. They are significant to say the least. They made my life worthwhile. Among other things I coached Little League Baseball with my sons as members of the team. I went to regattas in which they sailed, ran the regattas, scored them, organized them, and encouraged young sailors to participate. I spent 17 years trailering sailboats to these various regattas.

I had a 39-foot sailboat on which I sailed in many off-shore races. It broke my heart when, because of age and illness, I sold my Cal 39. My dream now is to buy a small boat, perhaps a 27-

footer, and go sailing again now that I am cured of much of my cancer.

I also led a full life with regards to contacts with my friends in various categories. One great friend was Admiral John Tierney, "the greatest pilot in the Navy," according to him. It's probably true. He commanded the Test Pilot Squadron. At one time before achieving flag rank, John had captained the Constellation aircraft carrier. He arranged for me to have a VIP cruise on the USS Theodore Roosevelt, a U.S. Navy Aircraft Carrier. We were going to land on board, but the carrier came into port, so we walked on board. It was a tremendous experience climaxing with me being catapulted from the carrier in a naval aircraft.

I must mention part of my career which was the happiest part. In 1959 I was asked by Dr. Ralph Stanton, one of the professors at the University of Toronto that I met during my graduate and undergraduate years there, who asked me to join the faculty at the University of Waterloo. He and a number of other professors that were dedicated to building this as the Canadian MIT. He offered me a post of Associate Professor of Mathematics. I went to Waterloo and began teaching that year and it. I loved the classroom. I loved working with young minds and feeling a sense of accomplishment when I could help them expand their horizons and capabilities. The following year Ralph promoted me to a full Professor of Mathematics.

I had also become quite friendly with Douglas Wright, the Dean of Engineering. In conversation with him, he suggested that I might want to move to the faculty of Engineering. He offered me the post of a Professor of Engineering and Director of the Department of Systems Engineering. This was an offer I couldn't resist. Once again this was exhilarating, rewarding, and exciting. I was given the PhD students. They were not much younger than I was. We often had games of touch football before going in for the seminar meeting. On one occasion I twisted my back. I led the seminar lying on the table in the center of the room. There was no precedent for this! It seemed to have set a first. It not only led to a reputation, as such, in the university, but also cemented a tremendous comradery with my students. They all went on to distinguished careers as department heads, senior positions with NASA, and professorships in distinguished universities.

To this day I am gratified that I had the ability for a short period of time to be a professor.

Chapter 5: CyberFone and Smartphone (Concept, Prototypes, Patents and People)

In this chapter, I intend to answer a few questions. What was the CyberFone? What did the patent cover? Why didn't it sweep the world before it? How did I come to invent it? What did I invent?

In 1994, I invented the Smartphone and called it the CyberFone.

I was appalled at the complexity of our systems at that time. We had as our clients the largest companies in the world. The systems we built for them were intimately connected with vital transactions utilizing communication networks on a worldwide basis and using communication networks at a myriad of local terminals. The computer programs were also ultra-complex. Telephones and their networks were in use as well as computers on a real time basis. Clerks, some wearing headphones, consummated transactions and then entered them into the computer. Others entered them directly into the computer at the same time. Some entry was done via voice. It was the 43rd year of the computer era but use of the computer was tedious and complex. We had accomplished a great deal, but more was needed. I sat down and invented the CyberFone. That diagram I drew is essentially what ended up on page 1 of the patent application, and later in the patent. This is shown in Appendix VII. That figure is also replicated within this chapter.

My son Paul suggested that I patent the design first. I did this. Then I proceeded to create and file the utility patent to cover the actual invention. The first of these I filed with the United States Patent and Trademark Office (USPTO) on May 19, 1995. Then I filed additional patents, listed in Sub-Appendix A of Appendix VI.

Fundamental in my original patent application was the use of touch screens and a camera. I wanted the CyberFone to be very simple and easy to use. While my first prototype was almost a cubic foot in size, I knew it was a factor of the technology of the time. I foresaw the day when CyberFone would be the size of a cigarette pack and would fit in a shirt pocket as it does today, ultimately reduced to the size of earplugs fitting in the ear. In the meantime, my prototypes became smaller and smaller in size, once again utilizing the advancements in technology towards miniaturization.

I was not satisfied with merely designing the hardware. Since I was creating a revolution and perhaps an industry, I decided to go all the way. I decided to simplify the complexity of creating new applications as well as the complexity of fitting existing applications into new computers. In our particular case, we were plagued having to use different machines and different operating systems in different parts of the world. I recognized that these were two separate efforts.

As I progressed in my parallel developments of both the hardware and the software, I came to the conclusion that I should combine the two into one patent. Hence my patent application covered not only the CyberFone, but also included reference to a more simplified approach to creating applications for it. My whole objective was to provide ease of use, mobility, unlimited use of the Internet, and new applications that would be independent of the manufacturer of the CyberFone. I foresaw the day when diverse manufacturers would license the patent for their own Smartphones.

Making applications completely independent of the hardware had fascinated me since the earliest days of the computer.

The operating system is that program which sits between the computer hardware and the application that permits the application to utilize the various components of the hardware without having to incorporate that particular code into the application structure. In computer parlance, we refer to these as drivers. Every device has a driver associated with it. This is a code grouping or application per se that is called upon by anything that a program requires. For instance, the screen has a driver. Any input unit has a driver. The modems have drivers. The operating system not only allows connection into the various drivers, but also keeps track of where you are in your program and where you are in the use of the various devices that make up the computer. In the mid-1960s we decided to try to get more utility from our programs by creating multi-programming capability. This in turn made the operating system very complex since that was where we brought the capability of having different programs with different positions in their path utilizing all of the components of the computer system to maximize the use of the machine in real time. It is obvious that operating systems are very complex, occupy a large part of space within the computer, and take up a significant amount of time and resource use on their own. At one time it was estimated that the operating system took half of the computer's available resource capability. That has been reduced dramatically today but is still a significant resource user.

One of the great problems associated with global systems is that you encounter various types of machines, each of which has its own operating system. This leads to costly maintenance procedures, and often to misadventure when the additions for a changed operating system do not function properly.

Our systems were critical. They monitored and controlled the flow of all financial assets in the corporation. In our case, our clients included the largest corporations in the world. Their operations were complex to say the least. Many of these systems were worldwide. Not only was the cash flow electronically controlled, moved, and documented, but investment, debt, and future cash requirements were all simultaneously coordinated. We could not afford errors. Hence, we spent a significant amount of time and effort at repeatedly testing before we incorporated any changes whatsoever in any of our systems.

I thought that while I was turning the entire industry on its head by combining telephones and computers into one instrument, I decided to take another step and eliminate the need for the dependency upon the operating system any time an application was changed or a new one added. I wanted to make the application completely independent of an operating system so it could function on any computer.

All of these features were incorporated into my patent application.

I was well on the way towards finally accomplishing my dream of having a computer generate the applications itself, rather than have this become a tedious ongoing effort on the part of highly skilled programmers.

The invention essentially consists of what is shown in the diagram below. This diagram is taken from the face of the patent that was granted for the CyberFone. This patent number 5,805,876. It was filed on May 19, 1995 and granted on September 8, 1998.

This diagram should be studied carefully to see that the CyberFone is an all-embracing

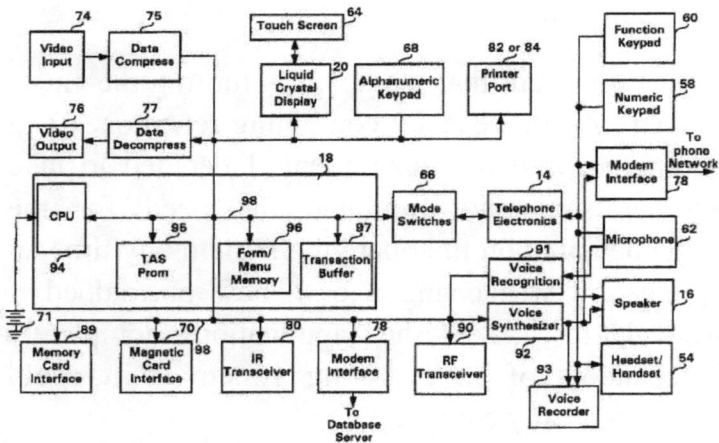

instrument that provides communication capability whether microwave or on telephone lines. It provides camera capability, it provides recording capability, and it provides access to databases externally. It combines the elements of a telephone and a computer into one basic instrument. It can be said the computer was placed into the telephone, or that the telephone as placed into the computer. I couldn't get it across to the patent attorney that what I had created was a combined capability so that the device I was using gave me access to the total world of the Internet, and to the total world of the telephone networks. This was not a case of making the telephone a data entry capability for a computer but rather providing all of the computer capabilities and communication capabilities in one place at one time. As I said repeatedly to the patent attorney, "I've put the world in your hands. That is the essence of the CyberFone. A universal communication device linking anyone anywhere anytime to the database with the capability of the internet, as well as the capability of the communication networks, whether wired or wireless."

With reference to the diagram, note the section that shows modem interface, study the diagram carefully you will see that it includes a headset, microphone, speaker, touch screen – in a sense everything you can have in a modern Smartphone as provided by the communication companies and others in this year 2019. That was the CyberFone.

I put more into the patent than just the device; it included 44 claims. I also added techniques for generating applications, and for operating the system with its own built-in operating system which I called TAS, for Total Application System. It was a complete application and, in my opinion, as will be outlined by Joseph Martino and Joseph Looby in the book *The CyberFone: The First Smartphone*, a revolutionary approach to the use of the Internet. TAS became a reality. However, it happened through piracy and infringement of my patent.

Below is the Abstract from the first page of the original Patent Application, depicting everything that the device (the CyberFone) would be capable of doing:

ABSTRACT

A data transaction processing system in which transaction data is entered by the user in response to prompts in a template which is tailored to each user application. The template and entered data are accumulated into data transactions which are immediately transmitted upon completion to an external database server for processing and storage. The data transactions are not locally stored for processing, and no conventional operating system is necessary. No local processing needs to be provided, and the only local storage is a flash PROM which stored the control firmware, a flash memory which stores the data streams making up the forms and menus, and a small RAM which operates as an input/output transaction buffer for storing the data streams of the template and the user replies to the prompts during assembly of a data transaction. The data transaction is received via standard protocols at a database server which, depending upon the application, stores the entire data transaction, explodes the data transaction to produce ancillary records which are then stored, and/or forwards the data transaction or some or all of the ancillary records to other database servers for updating other databases associated with those database servers. Also, in response to requests from the transaction entry device, the database server may return data streams for use in completing the fields in the data transaction or in presenting a menu on the display which was read in from the database server or a remote phone mail system. The transaction entry device is integrated with a telephone and is accessed via a touch screen, an optional keyboard, a magnetic card reader, voice entry, a modem, and the like.

We built a number of prototypes. The first of these, created out of balsa wood, I called the Woodbox. I contracted with a major corporation in our immediate area called CTDI, standing for Communication Test Development, Inc. CTDI was a premier communications development company. I went to them at the suggestion of Brian Kelly, at that time Executive vice President of Verizon. Their prototype was a cubic foot in size. We called the result "The Café." We programmed it and used the touch screen to verify the basic premise of our design. The next set of prototypes had all the functions and the camera which we considered vital to our design. They are still operational today. A complete description of the prototypes is included in the companion book, *The CyberFone: The First Smartphone*. Suffice to say by this time, June of 1999, we had a handheld prototype. It had most of the capability but lacked at that time the camera and the recording capability. These were in the other larger prototypes.

As a means of creating a rapid initiation for a handheld, I used a standard cell phone. The one I selected was that associated with T-Mobile, and we discovered that this telephone operated using a computer chip. We added our TAS software to the machine and found it possible to do all of the things we talked about in our patent except, as indicated, such things as the recording and the camera. These could have been put in, but we ran out of money and that's another story. And so, my concept of combining a telephony and computer capability into one instrument as outlined in the patent was demonstrated as early as 1996. As I used to say to people and to the patent attorney, "I put the world in your hands."

I decided that the computer product and the patent should be associated with different companies. Hence, I kept all the patents with a company which I ultimately labeled CyberFone Technologies, and I put the device manufacturing and ultimate sale into a company called CyberFone. The reason for this separation was that I had read and confirmed in discussion with our attorneys that in an extreme case where people who bought a device could also claim they also bought a license to the patent. By making the patent with a separate company with a license in turn to the product company, that risk was obviated.

We put together a tremendous Board. The Board consisted of (seen on the back cover of the Brochure in Appendix I): myself as Chairman; Vice Chairs were Da Hsuan Feng, PhD, and General P.X. Kelley, USMC (Ret.); Members included Michael G. Crofton, CEO/President of Philadelphia Trust Company; Rear Admiral Robert L. Ellis, Jr. USN (Ret.); Barbara D'Iorio Martino, LLD, PhD; Joseph A. Martino, MBA, MCSE; Joseph J. McLaughlin, Jr., CEO/President of Haverford Trust Company; Chris Pak, PhD, CEO/President of Molecular Targeting Techniques; David Shulkin, MD, Chief Medical Officer, Medical System of the University of Pennsylvania; and W. Kirk Wycoff, CEO/President Progress Bank.

Later we were joined by Frank Ianna, President of AT&T Networks. He became Vice Chairman, replacing Dr. Feng.

What is important to remember is the environment at the time. As late as 1995, cell phones were often in a satchel which was carried over the shoulder. They didn't reduce in size to being handheld until the late 1990s, approximately around 1997.

We were also going through the extension of the Internet to what we know now. Up until about 1995 it was almost a curiosity. It was used to transfer files, but the rise of the messaging and email capability was still in its infancy. The birth of handheld units and their widespread use were in their infancy and did not begin moving forward until well after the year 2000.

The CyberFone Impact

This invention dramatically simplifies the manner in which computer systems and applications are created. The end result is to dramatically simplify system creation (apps) and use. For example, at XRT we had some 1500 programs operating under five different operant systems and to a large extent, maintenance was expensive as well as a nightmare. With my approach, this could be reduced to one program with an unlimited number of parameter modules that would modify that one program to make it look like any application desired. The modern digital cell phone, Set-top box (video controller), and tablets all utilize the principles outlined in the CyberFone. The CyberFone is the Smartphone as we know it today. This was built and demonstrated in operation for more than 10 years before the iPhone and the iPad were launched.

I built the first CyberFone prototype which united telephony and the computer in 1995 and demonstrated its performance to the patent examiner and his supervisor in 1997. It utilized the technology of the day and appeared somewhat large and clumsy compared to the hand-held device of today. The bulk, however, was consistent with the bulk of cell phones at that time. It was not design but technology of miniaturization that is the difference. We demonstrated changes of face or application, as well as message transmission. I remember the supervisor saying to me something along the lines as follows: "If this were the 1950s, you would be coming to me with something akin to what the fax machine was then. You are claiming an ability to place multimedia capability in the hands of a user without the trappings of a main frame or a main frame operating system at hand." I said, "Yes!" I got the patent.

I organized a company called CyberFone Inc., to build and sell the machine, and organized CyberFone Technologies Inc. to hold the patents. I also organized CyberNet Inc. as a company to be parent of both of these companies. The effort was to separate the ownership of the patents from products and from license arrangements. Later, on legal advice, we did not make CyberNet a parent of the other two companies but maintained all three companies separate with CyberNet becoming the consulting organization. Later we changed the name to Cyber Technology Group. These gyrations of company names may be confusing, but this is to keep the record straight.

In Appendix I, I have reproduced the brochure we issued in June of 1999 in San Francisco at the Java One show. This brochure described the CyberFone process, and the pictorial in the centerfold told the story. During this meeting we were spotlighted by the executives of Sun Microsystems and won the "Best of Show" award. During that time, I think around June 19th, we also visited with various senior executives at Apple Computer Inc., including, I believe, Tim Cook, and started a working relationship with them. We were rejected. Tim told me that until Steve Jobs blessed a project it didn't happen.

What is important now in the material I reproduced (that was released at the Java One conference in June 1999), is the makeup of the board. I was proud of this group, especially Vice Chairman, Gen. P.X. Kelley USMC, a former

Commandant of the Marine Corps and member of the Joint Chiefs of Staff. He was very instrumental in arranging many meetings seeking financing and support for the CyberFone. He was repeatedly told there was no current market for such a product. It really amounted to the fact that we were 10 years too soon.

It took the marketing genius of Steve Jobs, plus an unlimited advertising budget, to launch the iPhone. If you will recall, he used the camera feature to gain initial interest in the iPhone.

During this time, I was very indebted to my son, Joseph Martino, who was the chief architect of the prototypes, working with me to hone my vision to the practical aspects of building the device. With him as Chief Technology Officer, we went on to build various forms of the CyberFone and demonstrated them at various shows. We won Best of Show awards in 1998, 1999, and 2000. We redesigned the machine in 2000 to make it more durable for our projected initial sales of 200,000 in 2001. The whole plan came apart when the chip supplied by one of the world's largest chip makers didn't live up to its specifications. Their own VP of Engineering laughed as he told us that the sales people exaggerated. We discovered it was deliberate oversell from their internal correspondence. We sued them, but finally settled out of court to come to a conclusion expeditiously.

The out-of-court settlement precludes me from identifying this company. But the damage was done. While the award was delayed until 2003, even if it had been achieved in 2000, the market disappeared when the DotCom bubble burst in 2000 and early 2001. There was no way that we were going to get any financial support for our machine and certainly there was no market even if the machine worked. We could get the telephone to work independent of the computer and vice versa, but when we put the two together, the transfer rate between the telephone and the computer would let you have a cup of coffee before anything happened. That's because the chip we had selected was operating at one tenth of the speed advertised even in the user manual. It was a design flaw that was uncovered because of the chipmaker's failure to supply the proper product to us. We had spent $10million on the CyberFone and had verbal orders exceeding 400,000 units. We foresaw a company doing a billion dollars a year in sales by 2002. Instead, we had a disaster on our hands. Our prototypes worked perfectly. It was our production models, using the faulty chip that were flawed. All attempts at regrouping became impossible. I did not have the additional two million dollars to redesign the machine around a new chip or the additional three million dollars required to finance the marketing and production cycle. Those sums had been spent on a production machine that too slowly transferred data between the computer and the telephone. We needed a new chip which would require significant design changes and would take another 12-15 months to make production units. My attempts at securing financing were difficult to say the least. In one aborted attempt, a man who had promised to put $500,000 into the company did not follow through. He wanted too much and would not even supply a little. By working hard, I had figured out how to overcome the design problem at a much lower cost.

Then I decided to begin looking at people who were infringing our patents and began work on this and initiated a series of letters in 2002, only to be felled by colon cancer. I recovered by early 2003 and developed a strategy in trying to monetize the patent asset. Easier said than done!

From 2003 until today, I have spent an average of quite a few millions of dollars in legal fees and efforts to monetize the patents. Throughout that period, I was helped in heart, spirit, and effort by one of the truest friends a man could ever have, Joe Looby. From the start as an important and vital employee, he rapidly became a friend who cared about the work, justice, and me. Joe has a particular knack for uncovering infringers, ideas for application for our patents, and new claims that, while implicit in the initial disclosure, are not explicit as claims. With his help, we progressed significantly in all aspects except in the legal breakthroughs that we sought but never seemed to accomplish. I sometimes wondered if our lawyers were more interested in fees than they were in our cause. I was appalled when, with Joe's help, we unearthed some of the outrageous padding in the legal bills that I had taken for granted as being proper in the past. What was a great anguish to me was that when we did find confident and fair lawyers, they soon had problems with conflicts in continuing to represent us. The strength and the weakness are that our patent claims are broad and cover so much of what is current today in communication technology. In my professional opinion, I knew the infringers, but I did not have significant legal expertise to prove it legally. Furthermore, the costs of infringement suits are staggering. For that reason, Joe Looby and I sought to widen the allowed claims for our disclosure to make litigation simple, and we also sought legal approaches to reduce the time and cost of litigation.

Along the way, a large number of additional patents have been secured using the original disclosure so that the priority date is always May 19, 1995. What we did was to make explicit in new claims what was implicit in the original disclosure. Just when it seemed that we were on the way to a series of successful legal engagements, I was felled by my first heart incident in 2007 and my hemorrhaging in 2003. In any event, we did file an interference action that would overcome the patents of a company that was generating $200million per year in royalties. Put side by side, the patents were identical except that ours were filed five months before theirs and was granted three months before theirs. This started the patent battle with the Patent Office that went on for years until we ran out of money. One heated argument that went on for two years and cost me $100,000 were the words "location" and "address". I used one of these and the other patent used the other. The patent examiner continually refused to accept the identity of these two words. His supervisor did. Finally, we had won. When I thought I was home free, the patent examiner pulled out a bulletin board application from 1990 and claimed that it was obvious from that patent to what I had invented. When I first got that letter in 2008, I didn't know whether to laugh or cry. Neither did my attorneys. The matter is absolutely ridiculous. There was no connection whatsoever. This too went through a series of appeals and actions until I ran out of money. We would have won that battle, but it would have been a pyrrhic victory. Such is the innovation life for the sole inventor.

As of this writing, I have negotiated a licensing of all of the patents to an organization that will sue infringers and remit to me a percentage of the gross proceeds. I have also retained a royalty-free license for use by a company I own, Cyber Technology Group. Here I expect to apply this technology to medical devices, especially incorporating Nano technology, and to new forms of communication, electrical illumination, and power generation.

But it may all come for naught. The millions spent on patents and development returned no "fruit." We almost won but the infringers took a different approach. They went to the White House. The Obama Administration backed a USPTO ruling that precluded non-users from exercising their patent rights. The patent law had been challenged by the accused infringers. They had mounted an attack to curtail the "nuisance" of the patent holders in threatening accused infringers with legal action while offering a license settlement. If they didn't infringe there would be no suits.

In all of this, the major battles between the patent holders and the accused infringers, the inventor is forgotten and ignored. I could not find a major law firm to handle my patent battles because they had a client that might be infringing. The patent accumulators became known as "trolls," a derogatory term created by infringing defendants in order to hide infringement behind name calling. That is not to say that some patent accumulators did not play totally by the rules. In my opinion, both sides erred to some extent, but patents have been infringed upon, and the sole inventor has been forgotten. Hopefully the government will stop listening to lobbyists and start listening to the people they purportedly represent.

The CyberFone - Smartphone

The CyberFone may be seen all over the world today with other names on it. The most prominent names, of course, are iPhone, iPad, and Smartphone. That is what I invented. Steve Jobs used the concept to make Apple the most valuable company in the world. He was a marketing genius. He never claimed to have invented the Smartphone but rather to have created a demand for it. Had I had a partner with that kind of capability, then perhaps CyberFone would have been the most valuable company in the world. I have included a biographical sketch written on the day Steve Jobs passed away on October 5, 2011. And yet, despite his great genius, he could not make the Newton (the first personal digital assistant) or Lisa (the first extended use PC) successful. They too were too soon for the marketplace. It was these failures that led to Vincent Sculley becoming CEO of Apple and with Steve Jobs being put out to pasture (fired) in 1985. This led to his work and outstanding success with NeXT and Pixar which he sold to Disney. His success with Pixar provided him with the bulk of his accumulated wealth. More importantly, his experience with Pixar and with the Disney Company honed a sense of market acceptance, penetration, and timing. I personally believe that that set of experiences were extremely useful and quite instrumental in conditioning the steps he took to make the iPod, iPhone, and iPad successful by stimulating the demand of the people of the world for these products. Once again, you will recall, he used picture-taking as a major capability of the iPhone.

I was 10 years ahead of my time and did not then have the perception of how to stimulate demand for a product that could become an outstanding success. However, I also didn't have the money. I was learning. I believe I could have overcome that obstacle without the bad luck of a bad chip. But such is life, and such is technology.

I often wonder what life would have been like had CyberFone become a billion-dollar company early in the 2000's. But this is musing on a hazy, lazy summer day, castle building in the sky, idle thinking, and daydreaming. It didn't happen. I regretted it at the time but do not now. The failure to build a working CyberFone in mass production gave me more time to think, write, play with my grandchildren, and survive challenging health issues over the years. I did not and I do not envy Steve Jobs. He was a remarkable man with a remarkable gift. I admired the man and his understanding of the human psyche that searches for drama in everyday life. He provided it with his form of presentation, with the genius he exercised in the products his organization produced, and with the marketing approach which aimed not at acceptance but rather to create a great desire to acquire. It was almost as if to buy one of these products was the ultimate accomplishment for the buyer. I would not trade places with Steve Jobs for anything, especially since I am still alive. The important thing is that I survived to be able to write this, to play with my grandchildren, and to live life to the fullest. I have no regrets.

*　　*　　*

In the first part of this chapter, I talked about a company that was the rage in email. That company was Blackberry.

While there on the West Coast I arranged to go to Apple. I had asked for a meeting with Steve Jobs. When I got there, I was told that Steve was too busy to see us, but we could meet with his immediate assistant. This turned out to be a gentleman I believe was Tim Cook, and two women whose names I forget. Tim listened very attentively to my presentation. I asked that Apple execute a licensing agreement and a marketing agreement. He said that it sounded reasonable, but until Steve reviewed it and made a decision to proceed, nothing could be signed. This was June of 1999. Please note that

Steve blessed the idea in 2007 when the iPhone was launched. There was very little difference between the iPhone and the patents I held. In particular, no one had put a camera into a cellphone until I did with my patent. Why is this important? Because when Apple began to market the iPhone, they did it by demonstrating the ease with which pictures could be taken. Many of the commercials showed people standing on the corner and taking a photograph. It was the photographic capability of the instrument that was the entrée into the battle of displacing the Blackberry and creating a market for the Smartphone. Steve Jobs and Apple had the smarts, and the money, to support an innovative marketing effort for this innovative instrument. The rest is known. Apple is the now the largest value company in the world, has a war chest exceeding $250billion in cash, and dominates the iPhone and iPad market. Their success has created many billionaires.

We went to a large number of firms and demonstrated our capability. One law firm we used in Philadelphia, Cozen & O'Connor, where the managing partner, Pat O'Connor, was a great personal friend. We tried to get them to consider the use of the CyberFone to replace the use of their Blackberrys. We were turned down cold. We were told there was no interest or advantage in being a pioneer.

We had a contract with the Coast Guard to develop systems for communication and security. We did not get to phase two of the contract. Some other firm that was using Blackberrys got that.

Jack McMakin, CEO of the Crozer Keystone Health System arranged for a demonstration to his financial people. Jack was very enthusiastic about what we were demonstrating, but we got no buyers. We were too revolutionary, and too soon.

In the hope of getting government contracts, I asked Congressman Curt Weldon, who was absolutely intrigued with the device, to set a meeting with Newt Gingrich, who was Speaker of the House at the time. I took the CyberFone into the House, into the office of the Speaker, and demonstrated. He was enthralled. He suggested that we try to get government research contracts and application contracts since he could see very far-reaching capabilities for this instrument.

Da Hsuan Feng, a Professor of Quantum Theory at Drexel University, and a supporter of Curt Weldon in his hopes for computer networks, became enthralled with the device. So did Dr. Chris Pak, who was President of a molecular targeting company for developing products for cancer control. Dr. David Shulkin, who was the Chief Medical Officer of the Philadelphia Hospital Networks, arranged for me to meet with Dr. Bill Kelly, who was president. Dr. Shulkin wanted to put CyberFones throughout the Philadelphia Hospital Network and Kelly agreed. It became a case of developing manufacturing capability which I will talk about next.

Up until this point in 2000 we were demonstrating the handheld unit that I had cobbled together from a T-Mobile cellphone. It didn't have the camera, but it had everything else. Perhaps the most enthusiastic supporter I had was Daniel Hilferty, President of Blue Cross. He could see tremendous application for the device.

I had a contract with the Norristown Board of Education to supply 50 units as part of the application of the CyberFone in education. These would be manufactured by the PEMSTAR Organization in Los Angeles, California, at a fixed price. We had an order in for 350 units.

PEMSTAR failed in giving us a unit that was consistent in use. Our contract with the Norristown School Board, which was a sub-contract of a larger contract with SAIC, the largest contractor in the defense area with the United States government. The contract was canceled, and we had to pay a penalty of $50,000 to SAIC.

What is interesting as a side light is that we had a separate contract with a division or a co-partner of SAIC, Telcordia. This was the former

Western Union company that was now owned by SAIC, and renamed Telcordia Technologies, Inc. They concentrated on developing and making products for the telephone industry. Our contract with them was to generate a means of rapid linkage of the phone networks and the computer networks within the CyberFone.

The most important contact we had was with a pharmaceutical supply company. They were absolutely intrigued. We had many meetings with many organizations and every meeting came away with enthusiastic supporters. They backed their enthusiasm with a verbal order for 400,000 units.

In an effort to guarantee a good product, I negotiated with a group of people in Buffalo to manufacture the device and enhance the design. When they produced the prototype, the performance was miserable. You could almost have a cup of coffee from trying to get the telephone to work in the middle of a computer application. This was late 2000. It turned out we had depended on the use of data transfer chips made by a major Japanese chip maker. They had advertised a minimal data transfer rate of 32 frames per second (full motion video) and in reality, we were only achieving a rate of 8 frames per second which resulted in the delays. We talked to their Vice President of Engineering and he sort of laughed and said the marketing people had gone overboard and he sent us a letter saying they were advertising a rate of 32fps when they were only achieving 8fps. So, they were advertising a rate four times the actual value. This gave us the perfect evidence for a lawsuit. We embarked on this and engaged in a prolonged discovery process that went on for two years. We finally got a hearing before an arbitrator and my lawyer was convinced to settle for $750,000 instead of the $10 million which we needed. I used the $750,000 to pay off debts that we had, but certainly not debts that I personally had. I will go into this in more detail in the next chapter. Suffice to say that we were destroyed as a manufacturing company when our product did not work at the end of 2000 and the telecommunication industry collapsed. You couldn't give anything anyway from 2000-2003 or 2004. We were out of business.

Our disaster was further impacted when I was in a rear-ended car accident and suffered a concussion and whiplash. I remember trying to run a board meeting in July of 2000 following the accident on my birthday in June. I recall distinctly that the faces seemed to weave in and out and I had a hard time remembering names of people I had known for years.

We regrouped and I decided we had to license our product vis-à-vis my meeting with Apple in 1999.

Inventor's Digest Cover Story

I was the cover story of the March 2014 issue of Inventors Digest. The article and photos follow:

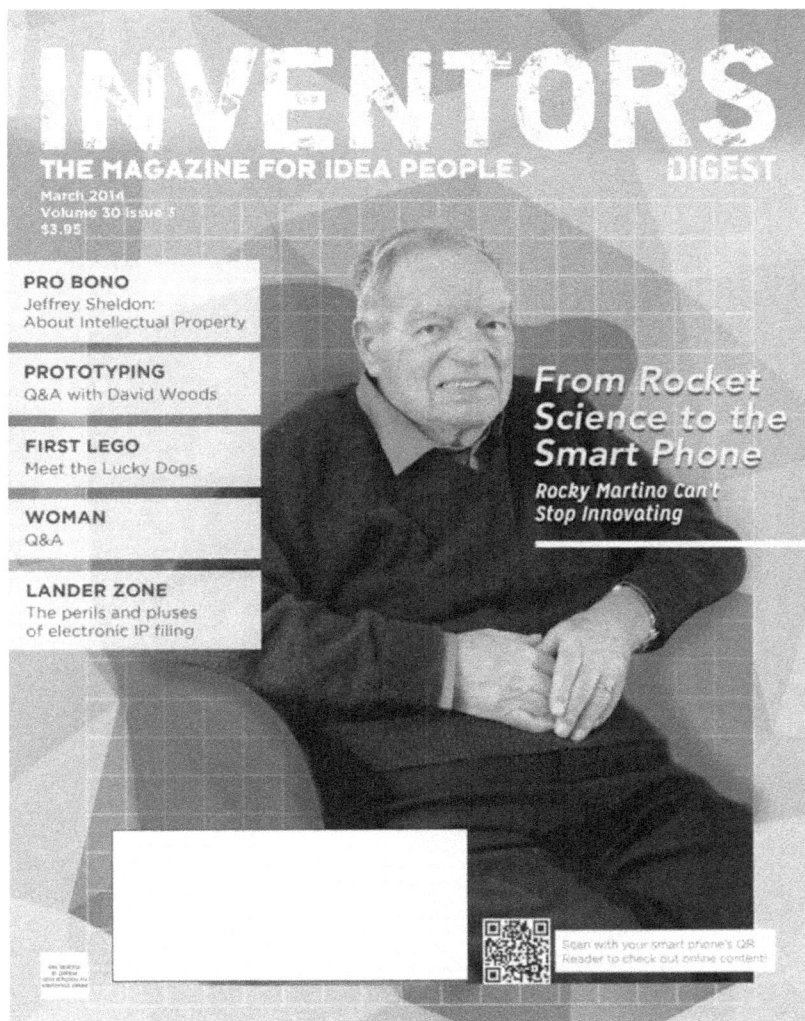

INVENTORS DIGEST

THE MAGAZINE FOR IDEA PEOPLE >

March 2014
Volume 30 Issue 3
$3.95

PRO BONO
Jeffrey Sheldon:
About Intellectual Property

PROTOTYPING
Q&A with David Woods

FIRST LEGO
Meet the Lucky Dogs

WOMAN
Q&A

LANDER ZONE
The perils and pluses
of electronic IP filing

From Rocket Science to the Smart Phone

Rocky Martino Can't Stop Innovating

Scan with your smart phone's QR Reader to check out online content!

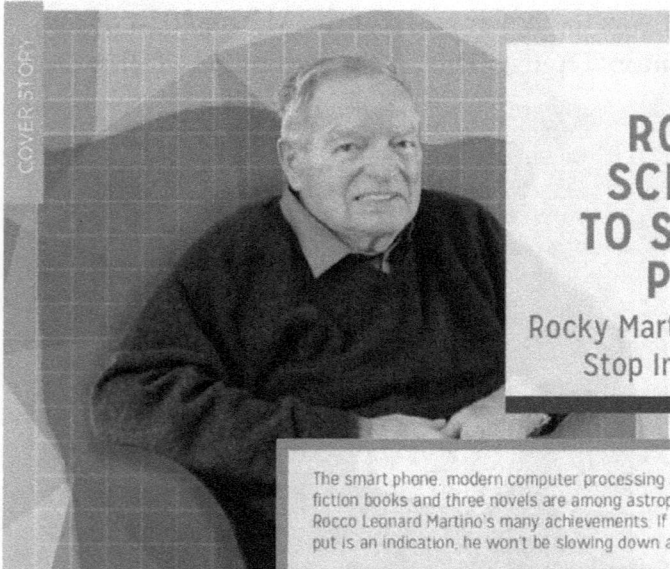

FROM ROCKET SCIENCE TO SMART PHONE

Rocky Martino Can't Stop Innovating

The smart phone, modern computer processing and 21 non-fiction books and three novels are among astrophysicist Dr. Rocco Leonard Martino's many achievements. If his recent output is an indication, he won't be slowing down anytime soon.

In the early 1990s when mobile phones looked more like portable breadboxes than communications devices, Dr. Rocco Leonard Martino, a trained rocket scientist who worked on heat shields using the earliest computers, was thinking of a better alternative. He reasoned that with the right chip set and innovative software, telephony and computer science could be united to produce what we call the "Smart Phone" today. He called it the CyberFone.

He wrote articles and gave speeches and demonstrations describing life using the CyberFone. He was at ten years ahead of his time. There were no effectively "intelligent" phones at the time because the size and cost of processing hardware limited data transmission and handset displays.

Dr. Martino is an astrophysicist and is frequently referred to as a rocket scientist and computer pioneer who worked in the forefront of space flight. His work began in 1951 and continued until starting company,

XRT, in 1972. His early positions included working with Sir Robert Watson-Watt, the inventor of Radar, and Dr. John Mauchly, the co-inventor of ENIAC. Today he is responsible for numerous innovations and patented inventions, many in the data processing and computer transaction fields. Around 1990, his thoughts turned to a challenge that he felt strongly enough about to invest in personally. If phones could be integrated with displays, with the proper hardware and software, users could meet their telephone and computer data access needs with a single device, the CyberFone.

Dr. Martino, 84, is President of CyberFone Technologies, Inc., which sold its patents in 2011 to CyberFone Systems, a firm acquired in 2013 by Marathon Patent Group (OTCBB: MARA), a leading intellectual property services and licensing company. Dr. Martino attributes his broad background in aerospace and interest in data processing as fundamental to his problem-solving prowess.

112

STAIRWAY TO THE SMARTPHONE

The foundational patent portfolio acquired by Marathon from CyberFone includes claims that give the holder the right to practice specific transactional data processing, telecommunications, network and database inventions, including those covering many financial transactions. The portfolio, which currently has an established base 40 licensees, consists of ten United States patents and 27 foreign ones, and one patent pending. The foreign patents include coverage in nine countries, including those in North America, South America, Europe, Pacific Rim and Asia.

Patents that cover digital communications and data transaction processing are integral to many applications in the wireless, telecommunications, financial and other industries. The CyberFone portfolio has been cited in patents owned by IBM, Cisco, Hitachi, Siemens, NEC and in 437 other patents issued or pending in the United States.

The portfolio includes patents whose claims cover processes for a telecommunications system that can be used in a menu-like format, allowing navigation and data input creating "data transactions," which are then transmitted to a database. Also covered are methods that detail the ability to input data in an "agnostic" form-driven operating system in which data can be processed and returned in real time.

Marathon believes the rights to CyberFone's assets are especially valuable in today's rapidly growing mobile-Internet environment and will provide licensees significant value. The priority dates of the patents are as early as 1995.

EMPOWERING PHONES

The smart phone concept was something that Dr. Martino started working on, in 1994, shortly after the cellphones, large and clumsy by today's standards, began being mass-produced. He filed his first patent in 1995, and the patent was issued by the United States Patent and Trademark Office in 1998 as number 5,805,676, "Telephone-transaction entry device and system for entering transaction data into databases." The Java One 1999 trade show named the CyberFone among the Best in Show devices.

An international authority on space flight, computer systems and digital financial transactions, Dr. Martino was trained in astrophysics at the University of Toronto with grants from the United States Air Force, the United States Navy, and the Canadian Defence Research Board for investigations. His research focused on the re-entry of space vehicles from outer space. This work was instrumental in creating heat shields for spacecraft. Not only is the smart phone attributed to Dr. Martino, but he also is one of the chief architects of the computer age and computer security. He has served many of the world's largest businesses, leading departments at Olin Mathieson, Booz Allen, Unisys, Mauchly Associates and has advised national governments and major corporations. With the company he formed in 1972, XRT, he created systems for international banking, trading, and debt systems for many of the largest companies in the world. By 1997, his XRT systems were transferring some three trillion dollars every day in international commerce.

Dr. Martino, who is known as "Rocky," is a dedicated "problem identifier," the first step in problem-solving. In addition to the 42 patents currently attributed to him, Dr. Martino has written many seminal technical treatises. He is responsible for an impressive list of books, including 21 works of non-fiction, "Ground Effect of Radio Wave Propagation" (1956) to "Creating the Cyber Age" (2014), three novels and a play.

Dr. Martino says that he has used his knowledge of the inner workings of government and business, together with his ability to identify and solve problems," along with a sense of humor," to find the answers to many challenging questions as well as to help frame a story.

EARLY YEARS

Dr. Martino's own story is as compelling as any books he has written. He grew up in Toronto, Canada through the depression years of the 1930s, building model airplanes with his brother and dreaming that one day man would fly to the moon. His father, a master chef, culinary judge, and author of two books on cooking, set the tone for creativity in the Martino household. "Father was one of only eleven members of the Epicurean Circle in London," recalls Dr. Martino. "Solving problems and exploring new ideas was his idea of good dinner conversation."

Space exploration was Dr. Martino's siren call in his early years, and in 1951 he was engaged in graduate work on early heat shields necessary for the re-entry of space vehicles, using one of the first computers in the world for his work.

DR. MARTINO HAS BEEN IN THE FORE-FRONT OF COMPUTER APPLICATIONS AND PROCESS INNOVATION FROM INCEPTION...

Prototypes of CyberFone devices

of the modern computer. His contributions have created products and jobs, promoted economic growth, and enhanced the ability of technology to address important aspects of business, commerce, and government, as well as to improve the quality of life.

His accomplishments encompass many aspects of product and process innovation, especially in four major areas. From 1951 onwards he was associated with procedures, techniques, and processes for creating actual computer code by the computer, without having the programmer or system designer write it. Simple statements in the vernacular and with formulae are created by the system program generator. These statements, much simpler than computer code, are then processed through a computer program of various types: compilers, assemblers, interpreters, translators, and system generators. Dr. Martino pioneered the concepts of automatics system generation from the 1950's.

His efforts in language formulation, compilers, translators and system generators long before such procedures became commonplace. Some of this work was in association with Dr. Grace Murray Hopper and the creation of COBOL. These techniques were employed in science, engineering and finance, directed to aerospace, navigation systems, and the application of computers to finance and insurance. Some of these efforts were in association with Sir Robert Watson Watt, the inventor of Radar.

A second achievement dates from 1959 forward and is associated with procedures, techniques, and processes for planning and scheduling any kind of project, most notably new product development and innovation. Dr. Martino pioneered the application of the Critical Path Method to many types of projects, and especially to linking project costs to the financial and departmental structure of organizations. Dr. Martino also developed variations of these network techniques for the creation of complex systems, some with probability parameters. These inventions helped address the production and financial requirements associated with minimal inventory of finished goods and raw materials – a system which later was extended to "just-in-time" manufacturing by the Japanese and others.

His initial efforts were in association with Dr. John Mauchly, co-inventor of ENIAC, Electronic Numerical Integrator and Computer), the first electronic general-purpose computer. These techniques were successfully employed in projects associated with new products, processes, and structures.

Dr. Martino has been in the forefront of computer applications and process innovation from the inception

SECURE SYSTEMS FOR MULTIPLE USERS

The third major achievement of Dr. Martino's career

114

STAIRWAY TO THE SMARTPHONE

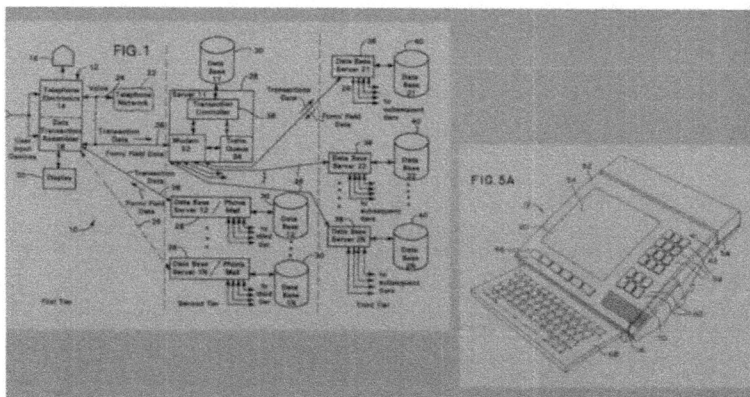

Figure drawings from U.S. Patent #5806676, Telephone/Transaction Entry Device and Sytem. Filed May 1995.

occurred from 1975 and is associated with procedures, techniques, and processes for creating secure systems for multiple users in networks both local and global. In particular, when applied to financial and medical systems, the need was for absolute security from penetration of any kind, for creating more than one copy of a transaction in real time, and for providing the capability for continuous operation in the event of any disaster, whether man-made or from nature. Dr. Martino's efforts were in conceiving and designing procedures for secure systems and for multi-user networks before others. By the mid 1990's, approximately three trillion dollars per day were processed through systems designed and created by Dr. Martino and his staff at XRT, a company that he sold in 1997 and which eventually became part of SunGard, a leader in data security and disaster recovery.

Most recently, Dr. Martino has been developing systems associated with convergent technology. In the early 90's, it became apparent that the next great advance in technology was the convergence of computers, telephony, and communication power into a single instrument that could operate in the hand or on the desk, in wireless or fixed mode, in networks of any kind, and would meet the needs of the user in varying circumstances without major costs of reprogramming and maintenance. All this capability had

to be simpler to use than existing PC's, hand-held devices, telephony networks systems and at a lower cost.

The work on this latest achievement lead to the creation of CyberFone, an early smart phone, as it continues today. Dr. Martino's efforts have been in conceptualization and reduction into patent-capable designs, processes and systems, followed by the creation and development of prototypes for proof of concept, and finally developing the capability for production and deployment. Dr. Martino was granted patents in the United States and in countries around the world. The most recent product of these breakthrough software and hardware concepts has been to provide these capabilities in cell phones, PDA's, and other hand-held devices. This application to wireless systems has created new capability with far reaching consequences for social interaction, emergency messages, real-time entry of data as events occur and instantaneous access to mass data.

His accomplishments have impacted the people of the United States and elsewhere, improving their quality of life and connection with others. The benefits to the U.S. and to global commerce have been significant and will continue in the years ahead.

"Dr. Martino is as much a renaissance man as any scientist the modern age has produced," says Doug Croxall, CEO of Marathon Patent Group, which acquired—

March 2014 | InventorsDigest.com 21

115

ROCCO LEONARD MARTINO

CyberFone last year. "Not only are his achievements impressive technically, they have real-world application and have begun to generate significant returns. He is remarkably humble about his accomplishments for so brilliant an innovator."

To date the CyberFone patent portfolio has generated more than $18 million in licensing fees, and the licensing program has been activated for less than three years.

SERVICE TO COMMUNITY

Dr. Martino is a deeply religious man, who has been knighted by the Vatican, and has served on the boards of numerous international organizations, including the Gregorian University Foundation, The Institute of Aerospace Studies of the University of Toronto, St. Joseph's University, the World Affairs Council, and the Foreign Policy Research Institute, of which he is also a Senior Fellow.

He has served as a Professor of Mathematics and as a Professor of Systems Engineering at the University of Waterloo and New York University. He has also lectured at other universities and institutes in North America, Europe, and Asia. Along with his corporate and creative activities, Dr. Martino an avid sailor, and has ocean raced for years. He is a Past Commodore of the Yacht Club of Sea Isle City and of the Mid-Atlantic Yacht Racing Association. He and his wife, Barbara, a trained research

> "MY ADVICE TO OTHER INVENTORS IS 'BE PERSISTENT,' AND MOST IMPORTANTLY, STICK TO THE OLD LATIN MOTTO - 'ILLEGITIMI NON CARBORUNDUM' - DON'T LET THE BASTARDS WEAR YOU DOWN."

Chemist, have established a charitable Foundation, and raised a fine family of four sons, thirteen grandchildren and one great grandchild.

Dr. Martino's contributions to science and philanthropy have been recognized with the Grant of his own Coat of Arms by the Governor General of Canada, four Knighthoods, Honorary Degrees from Neumann University, Chestnut Hill College, and Gonzaga University, and Lifetime Achievement Awards from the Chinese Monte Jade Society and the National Italian American Foundation.

"If you were to ask me 75 years ago when I was growing up in Canada if I would be responsible for all of this output, the inventions, the businesses, the books and the charitable work, as well as the family, I would have said 'not me'," muses Dr. Martino. "I simply gravitated toward what interested me and where I thought I could do some good."

When asked about the future, he laid it all out. "I am looking at new techniques for medical diagnoses, at Big Data applications, and immunology treatments for cancer. In addition, I fully intend to write at least one book per year as long as I can."

"As Machiavelli so aptly put it some five hundred years ago, 'Nothing is more perilous to success than a new system or idea: It will meet great resistance from those who are affected and only lukewarm support from those who will benefit.' There are plenty of people with an opinion about what is innovative and not, but listening to them won't do you much good. As a race we humans must innovate, not imitate, if we don't we will stagnate and eventually die. Our instinct for survival is like a compass that points us toward the future."

SOME OF DR. MARTINO'S CYBERFONE PATENTS:

US Patent No. 6,044,382
US Patent No. 5,805,676
US Patent No. 5,987,103
US Patent No. 8,019,060
US Patent No. 7,778,395
US Patent No. 7,324,024
US Patent No. 6,973,477
US Patent No. 6,574,314

116

Opportunities in Globalization
Published in the Philadelphia Inquirer June 4, 2006

Smart innovation is key to long-term prosperity. Nations won't thrive by just accepting low pay.

Rocco Leonard Martino is chairman and CEO of CyberFone Technologies; early in his career, Martino worked with the inventors of ENIAC, which celebrated its 60th anniversary in 2006

The seed of globalization germinated 60 years ago at the University of Pennsylvania, where ENIAC, the first operational general-purpose electronic digital computer, was unveiled in 1946. That single 11-ton machine has spawned more than 200 million computers worldwide, a number growing at 20 percent per year. Its stepchild, the cell phone, has grown to more than 1 billion units in one-quarter the time, and that number will at least double in the next 10 years. We can now connect with anyone, anywhere, at any time. This on-demand world fulfills the vision of a "global village" Marshall McLuhan put forward in 1964.

My argument here is that we need not fear globalization if we are ready to serve this rapidly growing world of on-demand markets. Such markets will be created and energized by Web-based labor, and our political and business leaders need to do all they can to encourage and reward it in the United States.

Web-based labor can encourage the rebirth of the American cottage industry as a potent economic contributor. It will help individuals and companies to conceive, make, sell and ship goods in low-capital-requirement environments via the Internet.

Globalization has accelerated dramatically over the last decade, leading to wealth redistribution and to dislocations here and abroad. Many Americans are concerned that the newly flat playing field may put us at a disadvantage in global commerce and trade. Nations whose workers will work for the lowest bidder seem to have an advantage, and larger, high-labor-cost nations such as the United States seem to lose out.

Thus, there are efforts in the West to stem globalization. In France, measures are being taken to prevent foreign acquisition of majority interest in a number of companies. Protectionist rules in agriculture have long applied in Europe, despite EU treaties seeking a level playing field.

But long-term benefits can counter globalization's short-range dislocations. Politicians too often blur the distinction between the two in seeking votes. To be sure, globalization seems to favor low-cost-labor nations. But the ultimate goal is sustainable, growing national wealth, not dominance as a lowest-cost manufacturer. Hence, the American economy need not suffer if it can continue to exercise profit control through innovation and invention.

Of the world's 6.5billion people, it is estimated that more than 1 billion use the Web. The power this gives people is the catalyst behind the "miracle" of Ireland; the burgeoning economies of China, Japan, and Singapore; and the software revolution and wealth creation in India and other nations. It permits us to buy a PC for less than $500, with the $100 PC on the way. Many complete service components, such as customer service, are now global.

The United States is well-positioned to compete in this new world. The A.T. Kearney global management consulting firm has developed a globalization index that measures countries' "economic, person-to-person, political, and technological integration." In its 2005 report, Singapore rose to first place, overtaking Ireland, which had led for the previous three years. The United States rose to fourth place largely on the growth in its Internet hosts and secure servers.

And in fact, high labor-cost, well-educated nations can compete with the low labor-cost countries, if they stay one step ahead. Progress lies with new skills, better management, new products, and above all, new industries. Globalization is creating forces that will demand new products and services geared to on-demand response.

Politically, our leaders can implement tax incentives for development and job creation in these new industries. In the workplace, the most far-reaching development will be telecommuting. The American workforce now lives in an information society - but it is stuck in an industrial workplace model. Energy consumption could be dramatically reduced with more telecommuting. The United States uses more than 20million barrels of oil a day, two-thirds of it in transportation. Sixty percent of it is imported. Consider the effect of reducing that by 20 percent, a target well within reach with telecommuting. The result would be to reduce pollution, travel time, auto accidents, and the overhead of conducting business, all while improving productivity and employee job satisfaction. The balance of payments would be reduced by more than $100billion per year, and GDP would increase. If we can outsource service work to Bangalore, India, then why can't we find a way to make outsourcing to American workers in Bangor, Maine, or Bowling Green, Ind., viable?

The "miracle" of Ireland can be our miracle - indeed, it can be every nation's. We simply need to engage in what has already happened with spirit and passion to succeed.

IP Watchdog - Op Ed Piece

I was asked to write an Op Ed piece of my experiences. It appeared in the May 1st, 2014 issue of "IP Watchdog" under the banner title ***Inventing the Smartphone: Why the 'Trolls' Were Saviors.*** Here is the text as it appeared.

Dr. Rocco Leonard Martino,
inventor of the CyberFone,
the first Smartphone.

A major car company is running commercials about companies that started in a garage. Amazon, Apple, Disney, Hewlett Packard and the Wright Brothers all started in a garage. The advertisement extols innovation, and the modest roots of great discovery and great companies. By implication, it hopes to align the public image of its cars coming out of a garage with these very successful companies. But the car company ad is wrong. It is highly unlikely these companies would ever get started in today's hostile environment for the small innovator or company.

Let me tell you why. This is a firsthand account of my experiences with inventing the Smartphone.

It became obvious to me in 1994 that the voice transmission over traditional telephone systems could be done using computers linked to telephone, cellular or internet networks. That was the origin of the concept of a Smartphone utilizing the computer in support of multimedia traffic. I called it the CyberFone, filed for patents in 1995, and built models during that same period to demonstrate to interested parties, including the patent office. I expected accolades for coming up with a useful idea, and proving it could be done, but that never happened. Some thought the screen was too small, others thought the concept of touch would never catch on, or that no one would want to make phone calls using a computer. One great business genius told me it was a software world and that no one was interested in hardware. That was a ridiculous statement. How could software run without a machine? Apple became the most valued corporation in the world by combining hardware and software.

It's tough trying to market a new product, no matter how exciting or useful it is. In fact, the more revolutionary the product is, the harder it is to gain acceptance. Disruptive technologies that lead to new industries are often delayed because they upset the status quo. It was Nicolo Machiavelli who enunciated this basic problem in his book "The Prince" in 1513. In today's idiomatic language, he stated, "There is nothing more likely to fail than a new system or invention, since it meets with the enmity of all those who are affected by it and receives only lukewarm support from all those who will benefit." The inventors of such disruptive technologies or devices are frequently ridiculed and often end up penniless. Johannes Gutenberg was met with tremendous resistance, even being accused of consorting with the devil, when he created his first printing press in the fifteenth century, even though his first book was the Bible. Rudolf Diesel died penniless; Frank Whittle had trouble meeting payments on his patent for the jet engine, which he filed in 1930. John Mauchly and Presper Eckert, who invented the first digital computer, the

ENIAC, in 1946, could not secure funding to monetize their great invention, and ultimately sold their company for a pittance.

In my case, I could get no financial support for my invention, which is why I set out to fund the project myself. It was a horrendous undertaking. Some obstacles included building prototypes, demonstrating to potential buyers, going to shows, extending the patents with new filings in different countries, writing programs to demonstrate the capability of my invention, and so on. When I was finally ready to go into production, the chip supplier falsified the actual performance specs. The chips I selected for the production versions only transferred data at ten percent of the rated speed, which was too slow. It was the year 2000. The whole thing, in a sense, blew up in my face. By the time I could recover from this disaster, the dot.com crash occurred, the orders evaporated, and I ran out of money.

Next, I tried to interest manufacturers in licenses. That went nowhere. One company officially told me no one would take me seriously until they were sued, so I set out to sue a few organizations that I felt were infringing. I could not find a law firm that would take the case. Firm after firm said they had a conflict of interest. This took years. As far as trying to manufacture, it seemed as though the train had left the station. What could I do? I could not get a major law firm to represent me in litigation because of client conflict, even when such firms were filing my patents. I could not go into manufacturing because I could not get funded. I could not sell licenses, even to infringers, because I had not sued them. This is when patent accumulators came into play. The opposition called them "trolls." To me, they were "saviors." They provide the experience and leverage individual inventors need to get the attention of those who take what they want without repercussion.

I turned to patent accumulation firm Marathon Patent Group (headed by Doug Croxall), who in turn engaged IPNav, (founded by Erich Spangenberg), and we struck a deal. Without Doug and Erich, my patents would have gone nowhere. Thanks to their efforts, along with the fiercely supportive legal teams at Russ August & Kabat and Mishcon de Reya New York, LLP, companies have recognized the value of my technology and acquired licenses. As an inventor, I am gratified at this success, but more importantly, I am heartened that the hard work and creativity has finally been recognized. It is unfortunate that justice can only be achieved with the application of the adversarial process of litigation. With the billions in profit achieved on Smartphones, I would have thought the inventor would have at least received a thank you instead of a brush off. Organizations like IPN and Marathon make it possible for the sole inventor to survive while swimming with sharks.

Though we started getting some return for the tremendous investments I had made in time and treasure with the CyberFone or the Smartphone, there was great resistance to paying royalties on my patents. At one court hearing, the defendants ridiculed my invention to cloud the issue of their infringement. Facts became irrelevant amidst the name calling, which seems to be the norm in this arena. There is a great hue and cry today about the patent trolls – a derogatory term applied to patent accumulators. But the trolls are not infringing. They are called trolls by infringers. When I was a little boy, we used to chant, "Sticks and stones may break my bones, but names will never hurt me." Maybe not in the real world, but the world of lobbying and high finance is a different story. There would be no 'trolls' if there were no infringers. The patent accumulators are providing a vehicle for small inventors to gain some return for their hard work in creating something new. Without such innovation, we would be stagnant. Innovation is crucial in progressing economically.

We need new industries to create jobs and opportunities for the millions of people all over the world who are having trouble finding employment. If we stifle innovation and prevent people from being creative and developing ideas, we will spiral downward. We must decide if we want the people of today to have a better future than their parents or grandparents.

If the government is sincere about stimulus to create jobs, then the government should set up a competition for new inventions and support what gets invented. A criminal is given the benefit of a Public Defender. An inventor is given nothing. Why not a government funded Inventor Defender? Criminals get help, but those who seek to benefit the country and the world get no help. But, you say, that's the SBA. Try getting an SBA loan for a new invention!

An even better idea would be for the government to honor the laws that protect inventors instead of changing the rules to protect companies.

*   *   *

<u>Closing Summary</u>

With such universal applicability and ease of implementation, the potential impact can be to create new industries. Using the Martino Technology can generate significant revenues, all protected by patented claims. This technology is in the forefront of the modern wave of convergent technology with a priority date of May 19, 1995.

I invented the Smartphone. I called it the CyberFone. It has changed the world.

The revolution happened!

a knight

in computing armor

Philadelphia Legate Rocco Martino is a renaissance man, and at 85 he shows absolutely no signs of slowing down

by Tim Drake

Without even realizing it, most of us use technology or computing processes created by Dr. Rocco Martino each and every day.

A member of Legatus' Philadelphia Chapter, the 85-year-old is also an international authority on finance and planning. Trained in astrophysics, he's also an expert on computer systems. In fact, he worked on the world's first computers and was instrumental in creating the technology behind the world's first smart phone. On top of that, Martino is also an author and novelist.

The sky's the limit

Growing up in Toronto in the 1950s, Martino dreamed big.

"I built model airplanes with my brother, and I dreamed of walking on the moon," said Martino. "I used to dream of space flight and how to do it."

Martino's dreams were fed by his excellent Catholic educators and role models. One of his favorite memories of growing up Catholic is the feast of Christ the King, which led Martino to dream of being a knight.

"In celebrating that feast we always had a boy chosen to portray a knight who would walk up the aisle of the Church," he recalled. "So as a young boy, I had this image of knighthood."

A lifelong Catholic, Martino studied mathematics and finance on scholarship at the University of Toronto, graduating *summa cum laude* in 1951. Still desiring to be a rocket scientist, he turned down opportunities with the Air Force and fellowships at Harvard and MIT to study astrophysics at the Institute of Aerospace Studies.

"The institute had contracts with the U.S. Air Force and Navy, so my doctoral work was sponsored by those agencies," Martino explained.

His first job involved setting the specifications for what would become heat shields, adopted by the National Advisory Committee for Aeronautics (NACA), the precursor to NASA. In order to perform the complex calculations, Martino used one of the first computers produced by Ferranti Electric.

"In one room was our office, in another room was the computer, and in another was the printer," Martino explained. "I wore roller skates to go from one room to another, performing calculations. I lay claim to being the first person to run a remote control printer by running strips of punch paper tape 110-feet long and skating to make sure the paper didn't jam."

Ahead of his time

Martino credits the creator of the first computer — John Mauchly — with his marriage to Barbara.

"During dinner with John Mauchly, I met Barbara," said Martino. "She was friends with his daughter. A year later, in 1961, we were married." The couple has been blessed with four sons, 13 grandchildren, and one great-grandchild.

Mauchly and Martino developed a partnership — Mauchly and Associates — and developed the Critical Path Method, an algorithm for scheduling complex projects. The planning technique was used in the creation of the submarine-based Polaris intercontinental ballistic missile and is still used in the research and design of large construction projects.

Throughout the 1960s, Martino headed up computer operations for other companies before launching his own in 1965 — R.L. Martino Company, which later became XRT, a company that created processes and systems to handle secure financial treasury

ROCCO LEONARD MARTINO

management. When Martino sold the company in 1997, it had 11,000 clients in 51 countries and was processing more than $5 trillion per day through systems he and his staff designed.

In the 1990s, Martino started building the first smart phone, known as the CyberFone. It's among roughly 60 patents that he holds.

Martino was at least 10 years ahead of his time. While he didn't have the money necessary to get businesses interested, the CyberFone patent portfolio has generated more than $18 million in licensing fees over the past three years.

"Dr. Martino is as much a renaissance man as any scientist the modern age has produced," Doug Croxall, CEO of the Marathon Patent Group, told InventorsDigest.com. The publication said that Martino's innovations "have impacted the people of the United States and elsewhere, improving their quality of life and connection with others."

Working for the Church

Through his personal and professional life, Martino's faith has remained his driving force, even in the face of challenges.

"In terms of faith, I went through some pretty tough studies," Martino said. "You can't study the origins of the universe and what holds it all together and say that it all came from chance. That's ridiculous. No real thinking scientist can claim that the universe came out of nothing. I propose that if you have an infinite source of energy, you can create an infinite source of mass. The infinite source of energy is God, and that source of energy created the mass that became the universe."

Of all his accomplishments — working on the heat shield, international finance, and the technology behind the world's first smart phone — Martino says his greatest accomplishments are found in his faith-based work.

Martino led the efforts to restore the country's first Catholic cathedral — Baltimore's Basilica of the National Shrine of the Assumption of the Blessed Virgin Mary. The archdiocese completed a 52-month, $34 million restoration project in 2006.

He's also proud of his work on the board of the Magnificat Foundation, creating days of spiritual enlightenment that have been held in major cities across the country.

"We had 3,500 people in a candlelight procession through the streets of Philadelphia," Martino said. "Public demonstrations and the witness to our faith are important."

As the author of more than 21 works of non-fiction and three novels, Martino says his novel *The Resurrection: A Criminal Investigation*, published in 2013, is his greatest accomplishment. Martino first became interested in the idea of novel writing after attending the Passion Play in Oberammergau, Germany.

"It had a tremendous impact on me," said Martino. "I still remember Mother Mary screaming, 'My son, my son, my son.' The play drove home that these were three-dimensional people with normal feelings of anguish and joy and pain. You don't get that when you just hear the scriptures."

Martino found novel writing as a way to bring the Scriptures to life.

"There are two great miracles related to Christianity," Martino said. "The first is the resurrection; without it there is no Christianity. The second miracle is the way in which Christianity swept through the Roman Empire. My novel is a means of telling that story."

The novel has garnered positive reviews. Thomas P. Sheahan wrote: "Martino has created a totally different view from the way the Gospel narrative is ordinarily presented. This book ... can serve as a springboard for a discussion of the historical circumstances surrounding the world where Jesus walked."

Now Martino is working on a follow-up novel.

"What you do to earn a living is not necessarily the most important thing you do," said Martino. "I feel I've accomplished something if I awaken in people an interest in faith."

Many of Martino's dreams have come true, including that longing for knighthood he had as a boy. Martino has been initiated as a Knight in the Order of St. Gregory, the Grand Cross of the Equestrian Order of the Holy Sepulchre of Jerusalem, the Order of Malta, the Constantinian Order of St. George, and the Order of St. Maurice and Lazarus.

"When I was initiated as a knight in the Order of Malta and the Order of the Holy Sepulchre, I kept thinking back to the way we celebrated the feast of Christ the King."

Dr. Rocco Martino has become a knight after all. **LM**

Dr. Rocco and Barbara Martino

The Martinos shake hands with Pope Francis on April 11, 2013

TIM DRAKE is Legatus magazine's editorial assistant.

CHAPTER 6: SALES EFFORT
AND LAWSUITS

Prior to going to the Java show in 1999 on the West Coast, we contracted a PR firm to develop a brochure for us. At that time, we considered ourselves CyberNet since to a large extent we were a network. The document showed the CyberFone at work. The displayed characteristics are important. Note that while the screen is larger, how, and what it is, is immediate. Ours was immediate. This brochure also shows our board at that time. We had a group of distinguished men and women:

Rocco L. Martino, PhD, DSc, PEng (Chairman)
– CEO, CyberNet Group, Inc.
Da Hsuan Feng, PhD (Vice Chairman)
– General Manager of HUBS, SAIC
Gen. P.X. Kelley, USMC Ret. (Vice Chairman)
– Partner, J.E. Lehman & Company
Michael G. Crofton – President & CEO,
Philadelphia Trust Company
Rear Admiral Robert L. Ellis, Jr. USN Ret.
– EVP & COO, CyberNet Group, Inc.
Barbara D'Iorio Martino, LLD, PhD
– President, LCA, Inc.
Joseph A. Martino, MBA, MCSE
– CTO, CyberNet Group, Inc.

Joseph J. McLaughlin, Jr. – President & CEO,
    Rittenhouse Trust Company
Chris Pak, PhD – President & CEO,
    Molecular Targeting Technologies
David Shulkin, MD – Chief Medical Officer &
    Chief Quality Officer, Medical Systems of
    the University of Pennsylvania
W. Kirk Wycoff – President & CEO,
    Progress Bank

This board met with many important organizations that dealt with structure. PX Kelley in particular was trying to raise significant funds to continue our research. I also worked with Charlie Dougherty, a retired congressman, to arrange possible government contracts. He was instrumental in setting up the meetings with Newt Gingrich, who was Speaker of the House at that time.

They were dedicated into moving the CyberFone forward. They were convinced that this device would sweep the industry. Things came to a screeching halt in June of 2000. The reason was the demise of interest in the Internet wonder-kinder, the total collapse of the stock market, and all Internet-related stocks. This depression of the telecommunication industry went on well into the decade beginning in 2000. It was resurrected to a large extent by the iPhone in 2007. But until then it was virtually impossible to get any capital for a start-up, and we were considered a start-up.

I had funded the entire operation personally. By late 2000 I was strapped.

The funds I derived from my sale of XRT were sadly depleted by a combination of gifts to my children, the creation of an educational fund for my grandchildren, and a donation of close a multi-million-dollar gift to Chestnut Hill College for the Barbara D'Iorio Martino Hall on their campus. After all I was left with about $5million dollars which I poured into the CyberFone. I was also working at $0 salary. The CyberFone expenses, patent fees, prototypes, travel cots, space rental, personnel costs, began accelerating. I borrowed the value of my portfolio and out ran that. I approached George Connell, Founder and Director of Haverford Trust, and Joseph McLaughlin, Chairman and CEO of Haverford Trust, and asked for a non-secured loan. They loaned me $1million. So here I was with loans exceeding $4million and only part of which were backed by my portfolio. I had to succeed with licensing. I made various arrangements with potential investors all to no avail. My suit against the chief infringer left me with only $500,000 after paying the legal fees. This was used to settle contracts I had with suppliers for the aborted manufacturing cycle. It was one disaster after another. Beginning in 2003, I made every effort to negotiate license agreements. I approached Verizon, AT&T, and RIM (Research in Motion) the company that owned Blackberry. They all turned me down.

By 2006, I was ready to throw in the towel when the iPhone was announced. In many of my discussions with major corporations in the country, when I mentioned my patents, the universal reaction seemed to be "well go ahead and sue us." So, I decided to do it. The trouble was I couldn't get a legal firm to represent me.

For example, the law firm I was using for my patents at that time Fish and Richardson, the largest such firm in the United States, could not represent me because one of their major clients was a target of any potential lawsuit.

My son, Paul, was a Partner at Alston & Bird LLP. I went to them. Their managing partner became intrigued and wanted to proceed but after further study they had a major client that would be a target of our lawsuits. With regret he turned me away.

My son, Peter had just graduated from Georgetown Law School and was employed by Quinn Emanuel Urquhart & Sullivan, LLP, a major litigation firm. I spent a significant amount of time with one of their senior partners Bruce Zisser, on the West Coast, who became interested. Ultimately, they turned us down. He did recommend a small litigation firm that I went to. They looked at it and turned it down. I never did find out which of their clients would compete.

I had been courted by patent accumulators, called "trolls" in derision, by the infringers, and got ridiculous offers from most of them. One organization, headed by Doug Croxall, whom I met previously and whom I came to like and respect, had a track record of success. We negotiated a contract that gave me an immediate $3million plus a percentage of downstream recoups. We immediately embarked on suits on 169 companies. 101 of these settled, and the other went to the White House in the Obama administration. The target, as they claimed, was the trolls who were interfering with their ability to innovate and created new products. They must have had somebody in PR write that for them. I was the inventor for God's sake. I was being infringed. I wasn't inhibiting their ability to function; they were inhibiting mine. What a role reversal! In any event they prevailed and managed to get the patent office to change the interpretation of the patent law so that only practitioners could use patent rights. Hence, I was now a litigant, not a practitioner, and had no rights to exercise my patents. Ridiculous.

In any event, the operations with Doug Croxall, who had taken over from Erich Spangenberg, and his company Marathon Securities, came to a halt. Subsequently they returned the patent rights to me. In any event the money I derived from all of these operations allowed me to reduce my loans at Haverford Trust from $4 million to zero. I am still left with other loans exceeding $1 million which I hope to repay in the near future.

And such is the story of all the attempts we made to derive some kind of monetary return for the $10 million and 10 years of labor that I had put into the effort.

The reader should recall that Frank Whittle who had invented the jet engine lost all rights to his patents when he couldn't pay his renewal fees. Rudolf Diesel, inventor of the diesel engine, died penniless and bankrupt. Sir Robert Watson-Watt, inventor of radar, had to proceed with action in the equivalent of the Supreme Court in England, the Privy Council, to secure his rights as the inventor of radar. Since this was done for government use, no royalty could be claimed. But at least he had the benefit of being named as the inventor of radar. John Mauchly and John Presper-Eckert, the inventors of ENIAC were denied their rights in a law case in Minnesota that favored the infringer, Honeywell, who was the largest employer in the state of Minnesota. The inventor, rarely if ever, benefits from the invention even if it changes the world, my invention did change the world. It created a trillion dollar a year industry. For this I was left with a $10 million debt.

Chapter 7: Summing Up

I led an eventful life. I changed the world forever with my work. There was never a dull moment. I thought it was over many times, and most especially this past Christmas. I am saddled with dialysis five days a week. I'm saddled with bone cancer. I'm saddled with the remnants of bladder cancer. I'm saddled with ongoing lymphoma, suppressed at this time. But yet I've never stopped. I won't let any of this stop me. Why?

I believe I have led a blessed and gifted life. I have survived where I have had no right to survive. At the age of 89 I got a severe flu type-A and pneumonia simultaneously a few days before Christmas. And my temperature soared to 103.7 which for someone with kidney failure is 104.7. Yet, within three days, I was fine. I was home for Christmas. Why?

Why do I survive and keep going? I think it is mutual. I think my constant drive has a way of jacking up my immune system, so I just keep going. I think I've taxed my body to the fullest, and my body responds. I also think I have been blessed beyond the average. And it has given me the chance to accomplish a great deal. Whether this is useful or not, I can say that I have satisfied my curiosity, I satisfied my ingenuity, my inventiveness, and my innovative streak to produce solutions and results in many diverse fields. These have ranged from Arctic Warfare, to space flight, to international commerce, and the development of devices, most especially the CyberFone. During my lifetime I have been present at the birth of space flight, the birth of the computer, and finally the birth of personal communication and data-search with the Smartphone. I didn't call this the Smartphone, but I did write articles about the Smart Society with the CyberFone. I'm not a word smith, but just a practitioner in the field of technology. Articles written about the CyberFone and Smart Society are in the appendices and mentioned throughout this book.

In the early 2000's Singapore was touted as the first and foremost wired city in the world. I proceeded to Singapore and tried to make an arrangement for the use of the CyberFone there. I was met with great receptivity, especially with my lectures at the University of Singapore. I wanted to proceed but didn't have the financing to do it. For that reason, I returned to the United States and embarked on the course of litigation. This I described earlier.

So, what did I invent? First and foremost, I put the computer into the telephone. This created a single instrument which put the world into the hands of the user. The abstract for my patents in a sense had it incomplete. It stressed only one aspect. The description, on the other hand, was totally correct. The figure on the front page of the patent says it all.

Why did I not succeed? First and foremost, I invented too much. Had I only secured a patent for a camera in a cell phone I would have made a fortune. I would have swept the market. But I put it all in the patent. Steve Jobs, on the other hand, was a better, infinitely better, marketing genius than I was. He used a single feature, the camera, in order to sell his concept of the iPhone. He also had the resources to implement a far-reaching advertising campaign of that capability. He also initiated an invitation to contractors to develop applications. On the other hand, I produced an App Creator system as well so that they could be done by a contractor or by the parent company. Once again, I didn't have the money to advertise that capability. We tried to replace the Blackberry. Jobs ignored it. We went after corporations. Jobs went after the people. He got to the people through the use of the camera, and the use of the iPhone mushroomed from there.

The lesson was learned, but too late. So, the question remains. Did I use my talents to the fullest? I should mention at this time I was driven all my life by what the nuns drove into me as a very young person. They kept hammering away, "To whom much is given, of him much shall be required." I took this to heart. I knew I had been given great gifts; I knew I had to use them. So, I did.

What was the objective of the use of my talents? Was it: job fulfillment, personal satisfaction, or being able to return in some measure what I was given?

Along with being a Professor of Mathematics and then a Professor of Engineering at University of Waterloo, I was also appointed a full Professor of Mathematics at New York University. Granted this was an Adjunct Position but it was a recognition of ability. While in New York I did lecture in the graduate school of NYU. I seemed to recall delivering a course in advance techniques of operations research. In any event that together with my experience at Waterloo were some of the happiness of my life.

Throughout my life, money was never important, but an essential ingredient in order to pay my bills. Recall for a minute the milieu of the time in which I lived. Companies were sprouting up all over, entrepreneurs worked in the knee hole of desks, and companies such as the nascent Microsoft soon became giants. I tried my hand with Mauchly Associates, Information Industries, and then XRT. In XRT in 1982 I was offered an opportunity to proceed to an IPO to raise $10 million for 25% of the company. We were in the midst of working on a GMAC contract and had caught the attention of Montgomery Securities. That together with AAA recommendations from GMAC set the stage for their offer. It was immediately canceled out when American Express launched a plagiarism lawsuit against us. Apparently, they got a copy of my source code and went to the patent office and registered it as their own and secured a copyright on it. They were suing me for plagiarism and theft. I looked back and found it online and it said "Copyright 1970 by XRT Inc." We proceeded to the copyright office and had the copyright canceled and our own instituted. Our case was over. We won. But the cost was $440,000 and it took every cent we had, every cent we could borrow. The total fees American Express had paid us was $72,000 for the work that we did which made them number one in the commercial paper industry. Their gratitude was to pursue this "nefarious suit," to quote an officer at Prudential Insurance who knew about this when it occurred.

That left a total sour taste in my mouth for anything approaching the idea of a public offering. Hence when it came time to sell XRT in 1997, we proceeded on a private route which led to our merger with Cerg Inc., a French firm listed on the Bourse in Paris. We also had an offer from Safeguard. In retrospect we made a mistake. We should have taken the Safeguard offer. But bygones are bygones and cannot be undone. Surprise, surprise, XRT-Cerg, replaced by the name XRT alone, was sold to Safeguard later.

In any event the sale or merger with Cerg gave me enough money to proceed with my CyberFone invention. It was enough money, as I recounted in my last chapter. If the Japanese chip had performed as advertised, perhaps the outcome would have been different. Such is life.

My son John has a way of saying, "You don't get out of life alive." Very interesting. Very direct. This is from a man who has come close to death many times. He sailed in a direct line from St. Thomas to Maine on a 110-foot schooner. Just think of it. The vast reaches of the ocean with nobody there.

So, what is the legacy I will leave? Well first of all I haven't left yet. I have no intention to stop continuing to write, develop products, and develop systems. As a matter of fact, I have an idea for two blockbuster systems, and a patent application already filed for a robot. So, I foresee further developments. For now, however, I think my basic legacy is that I have been gifted beyond measure, I have the right genes and immune system to insulate me from many of the perils of disease which often to lead to premature death. In all of this I believe I have been bolstered by the power of prayer. I'm a strong believer in prayer. Whether I would be considered a religious man or not is immaterial to me. I have rather strong beliefs. I think positively the fact that I was cured of bladder cancer by my trip to

Lourdes, France. I went there riddled with cancer that was invasive and on the verge of metastasizing throughout my body. I came back with no cancer in my body. I think it was prayer.

So, what is my legacy? I think at this time it must be the CyberFone. I have created something that has changed the world.

Appendices

Appendix I contains scans of the brochure distributed at the Java conference organized by Sun Microsystems that began in San Francisco on June 19, 1999. The CyberFone version we exhibited were meant to be used as point-of-sale devices, as well as desk-top units for business, education, industry, and commerce.

A close examination of the centerspread of this brochure shows videoconferencing. This was demonstrated at that conference.

Appendix II shows scans of the brochure associated with the prototype we exhibited, a CyberFone VIP 12. This unit was awarded a best of show award.

The VIP 12 was a touchscreen system. The handset was unnecessary and included a futuristic design to capture the attention of the people to make the move toward the Smart Phone. However, the phone was designed for the screen to be removed and used much as a tablet operates today. Everything was in the screen. With an adequate battery supply, it could be removed from its pedestal and carried about without the handset, just using the microphone in the body of the CyberFone. We, however, had a feature beyond the capability of the current tablet;

telephony. Of course, this is done today as an app. We had it built in.

Throughout the conference, we demonstrated videoconferencing on the three machines we had available. We also demonstrated multiple user text editing. In all of these operations, we demonstrated simultaneous telephony.

At this conference we aimed at the commercial market. We were working on the personal user market separately. There we were imparting links to the operating system of cell phones that had microchips turning them into CyberFones. As we demonstrated at the show, we each individually demonstrated with cell phones we carried throughout the convention center.

Appendix III contains diagrams of the CyberFone at work for individual consumers; in a local enterprise with legacy equipment; in virtual conferencing, distributed classrooms, and educational/ entertainment auditoriums and in communication at work in a large scale distributed networked enterprise.

Appendix IV has several panels which discuss the software components of the CyberFone system and the people behind the product.

Appendix V has a copy of the presentation describing the Uni-Link Virtual Application

Platform, which is essentially the application driven operating systems that current Smart Phones run on.

Appendix VI has the document titled The Martino IP Technology – A Historic Document of Capability. This elaborates on all the areas covered in the CyberFone patents and during development.

Appendix VII has the first page of United States Patent Number 5,805,676 – Granted on September 8, 1998; along with the URL where the entire patent can be viewed.

Appendix VIII is a brief biographical summary about Dr. Rocco Leonard Martino.

Appendix I
Brochure from 1999

<u>The CyberFone Brochure</u>

As described previously, we exhibited extensively at the Java Conference in June of 1999 in San Francisco. We displayed the CyberFone and won a Best of Show Award. As part of the material distributed, we produced a set of materials describing not only the CyberFone but how it worked.

Who We Are

The CyberNet™ Group, Inc., is a communication convergence company that produces software, network operating systems overlay and hardware products that empower the business or individual customer with the most cutting-edge communications available today.

In today's fast-paced world, CyberNet™'s applications are almost limitless. They include: business, education, healthcare, personal banking, hospitality, government/military, e-Commerce, e-Trade, and consumer. The CyberNet™ flagship device, CyberFone™, is a full-scale communication appliance, similar to a telephone, with interactive video, voice, data, e-mail and Internet capability, all accessible over a standard telephone line.

CyberNet™ began in 1995 as the vision of Dr. Rocco Leonard Martino, one of the original computer pioneers. Dr. Martino is one of the world's leading experts in the application of computers and communications in business, government, and industry.

Dr. Martino has been a leader in the technology community for almost 50 years and spent 25 years directing the treasury software giant XRT, Inc. His belief that it is possible to create an integrated system of communications capabilities that is simple, reliable, and affordable was the inspiration for the CyberNet™ Group, Inc. family of products.

Three years after establishing the patents, designing the products, and developing the software with his son Joseph Martino, Dr. Martino incorporated the CyberNet™ Group and brought on a seasoned management team led by Navy Rear Admiral (Ret.) Robert Ellis. This team shares Dr. Martino's vision and they are part of a close knit-group that has transformed CyberNet™ from a dream into a reality.

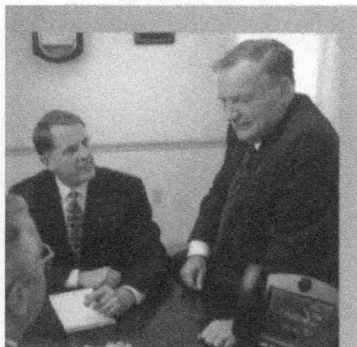

Today, with Dr. Martino (right) serving as Chairman and CEO, RADM. Ellis (middle) as Chief Operating Officer, and Joseph Martino as Chief Technology Officer, CyberNet™ Group, Inc. is poised to lead the next generation of voice, video and data technology by bringing communication to people and places that you never thought possible.

152

Convergent Communication Systems
Our Vision and Mission

Our vision is people to people – anywhere, anytime.

Our mission is simplicity – a convergent communication system for businesses, governments and individuals that is simple, reliable and affordable.

Today, knowledge is power and our goal is to provide those with a thirst for knowledge easy access to the expanding information superhighway.

People-to-people, CyberNet empowers everyone, from the CEO of a multinational corporation conducting a "cyber" virtual meeting with top executives, to the teacher assisting an array of students in the virtual classroom, to doctors making complex diagnoses from opposite ends of the country.

The CyberNet family of products allows all of us to see, listen, speak, and communicate with others as never before at a fraction of the cost of most personal computers. With the CyberNet Communications System, the convergence of voice, video conferencing, e-mail, and the World Wide Web is now possible for everyone simply.

Communication made better. The future made simple.
Simply the Future.

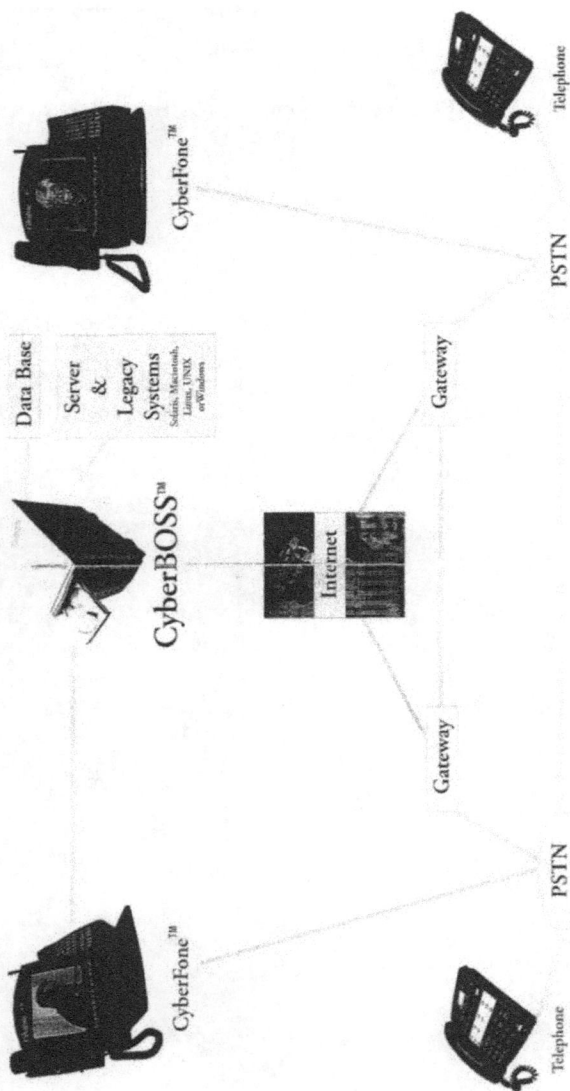

CyberNet™: People to People

Products and Applications

CyberNet™ Group, Inc. offers an array of communications convergence products - client software, network client server software and hardware - that are independent and easily allow users MacIntosh, Solaris, UNIX, Linux, SUN or Windows applications options.

Software

CyberNet™ Virtual Meeting System - The Virtual Meeting System is a stand-alone product that is also an integral part of the CyberFone™ unit. By itself, the "whiteboard" portion allows users to share and edit images and text in a paint program-type environment. Co-workers in separate offices can design business presentations together from different locations. The system also includes voice and video conferencing, text chat and file transfer.

CyberAGE™ (Application Generator Engine) - CyberAGE™ is thin client software for any application. It allows direct funds transfers, bill payments and all personal banking needs via connections to various legacy systems. Users can securely perform e-Commerce and store transactions, and move data back and forth through CyberFone™. Like the Virtual Meeting System, CyberAGE™ can be run through a personal computer or the CyberFone™.

Network

CyberBOSS™ (Broad Operations Systems Server) - CyberBOSS™ is the control software that connects each source to all destinations - from laptops, CyberFones™, Apples, and network computers to Internet Services, legacy systems, data warehouse systems, and any desktop - and vice versa.

Hardware

CyberFone™ - The CyberFone™ unit operates under the CyberAGE™ software application and can be used individually, with a personal computer, or as part of a computer network. It provides interactive Video, voice over IP, Internet access and data transactions by simply plugging the unit into one regular electrical outlet and phone line. CyberFone™ comes with a wired handset and phone base, adjustable LCD display panel, retractable keyboard, small digital camera and hands-free microphone/speaker.

CyberNet™ Applications:

Business - Employees from different locations can collaborate on the same document or graphic simultaneously, bringing complex presentations together quicker and more efficiently. Large and small companies can conduct virtual meetings with participants dialing in from literally anywhere.

Education - Homebound, students in remote venues, and classroom students can all actively participate in a virtual classroom and receive instruction via their computer. Teachers can conduct lectures, class discussions, and class projects in this classroom without walls.

Healthcare - Voice over IP, video conferencing, and the Virtual Meeting System help doctors consult with colleagues on complex diagnoses, share information quickly and treat patients more effectively. In the doctor's office, CyberFone™ facilitates the waiting room check-in process with on-line access to individual patient medical records, including X-rays and CAT scans.

Hospitality - Hotels can offer guests a full range of communication options desired by both business and leisure travelers. Whether for last-minute reservations and confirmations, checking e-mail, receiving important documents, or making late changes to a big business proposal, CyberFone's flexibility keeps anyone on the move accessible.

Personal Banking - Pay bills, balance your checkbook, and perform all on-line banking tasks with CyberAGE™.

Government/Military - Military personnel stationed in faraway places can see and talk to their loved ones, making their deployment a little less lonely. CyberFone™ also allows for government-to-government connectivity at all levels.

E-Commerce - CyberFone™ puts the power of the Internet and the ease of e-Commerce at your fingertips. Purchase music, books, clothes - just about anything - through CyberFone™ software and hardware.

Consumer - For less money than many personal computers, you can not only talk to your friends and relatives across the country, you can also see and interact with them.

Appendix II

The CyberFone VIP 12

Cyber Technology Group

CyberFone

VIP 12

CONVERGENT COMMUNICATION SYSTEM

The Cyberfone VIP 12 incorporates web access, e-mail, video conferencing, and word processing, along with patented thin client software to seamlessly access corporate databases. This unit integrates voice (PSTN or VoIP), video, and data into one utility addressing both the need to access corporate data as well as the need to manage customer information.

VIP 12

SPECIFICATIONS & FEATURES

Operating System	Windows 98
Host Processor/CPU	333Mhz
Memory/Host	64 MB DRAM (Expandable to 128 MB)
Hard Drive	4 GB Drive
Display Screen (LCD)	12" Diagonal Dual Back Lit TFT VGA, 800 x 600, 16 Bit Color, Tilt Angle
System Navigation	Stylus Pen and Touch Screen (Mouse Optional)
Keyboard	Optional - Full QWERTY Keyboard (English) OS Enabled, Standard Notebook Size, Infrared Capable
Modem	56 Kbps V.90 Voice/Telephony
POTS Line	One-Line Operation, Full Duplex Hands-Free Speaker Phone
PBX	(Optional) Digital Emulation to Seamlessly Link to Most PBX Systems
I/O Ports	Parallel Port, Serial Port, 2 USB Ports, External VGA Port, RJ-11 Analog Phone Line Jack, RJ-45 10/100 Ethernet Jack, 2 PS/2 Ports, Microphone In, Speaker Out
Power Module	External AC Power Adapter, 90-260 VAC, 50/60 Hz DC to DC Converter for System
Telephone Dialer	18 Traditional Telephone Function Keys, Dialing Keys (0-9 Including # and *), Redial, Speaker Phone, Hold, Volume, and Flash
LED	Power, Speaker, Hold and Message Indicators
USB Camera (Optional)	Removable Base and Swivel Mount
Low Power Standby Mode	POTS Call Available
Universal Bay	Expansion Space Available for Mounting Bay

Full Tilt Angle, Touch Screen

Analog, PBX, IP Handset

Optional Magnetic or Bar Code Reader

USB Camera

OPTIONS

- Magnetic Card Reader, Track 1 & 2 or Track 1, 2, & 3
- Bar Code Card Reader
- Universal Bay for PCMCIA Interface, Floppy Disk Drive, or CD ROM

APPLICATIONS

- Hospitality
- Healthcare
- Government
- Education
- Financial/Banking
- Kiosks
- PC/Workstations
- "Sweep-the-Desktop"

CARD READER OPTIONS

- Identification Card Systems
- Access Control Systems
- Membership Management Systems
- Security System
- POS Systems
- ID Card Applications

Cyber Technology Group

989 Old Eagle School Road, Suite 800
Wayne, PA 19087
(610) 989-9330 Fax: (610) 989-9366
www.cyberfone.com

CyberFone

© 2000 Cyber Technology Group, All Rights Reserved Worldwide

Windows 98/2000 Operating System

Appendix III

The CyberFone at Work

CyberFone for the Individual Consumer

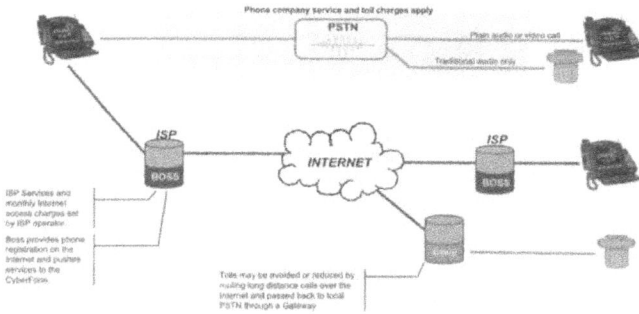

CyberFone Communication Systems
in a Local Enterprise With Legacy Equipment

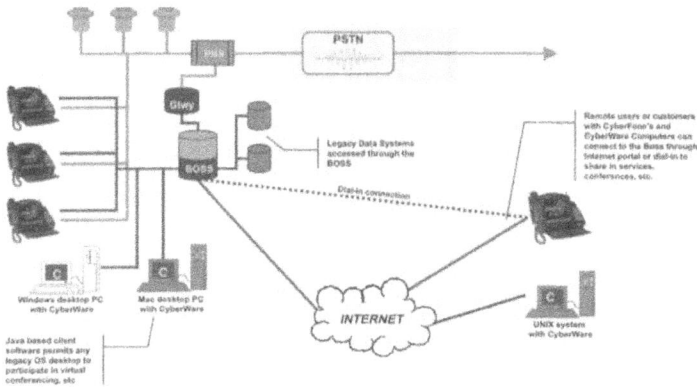

**CyberFone Communications Systems
Enable Virtual Conferencing, Distributed Classrooms,
and Educational/Entertainment Auditoriums**

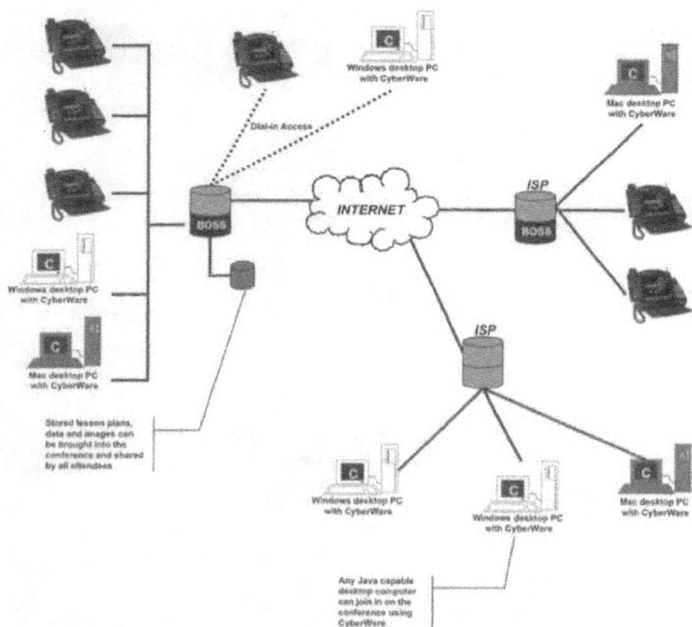

**CyberFone Communication Systems in a
Large Scale Distributed Networked Enterprise**

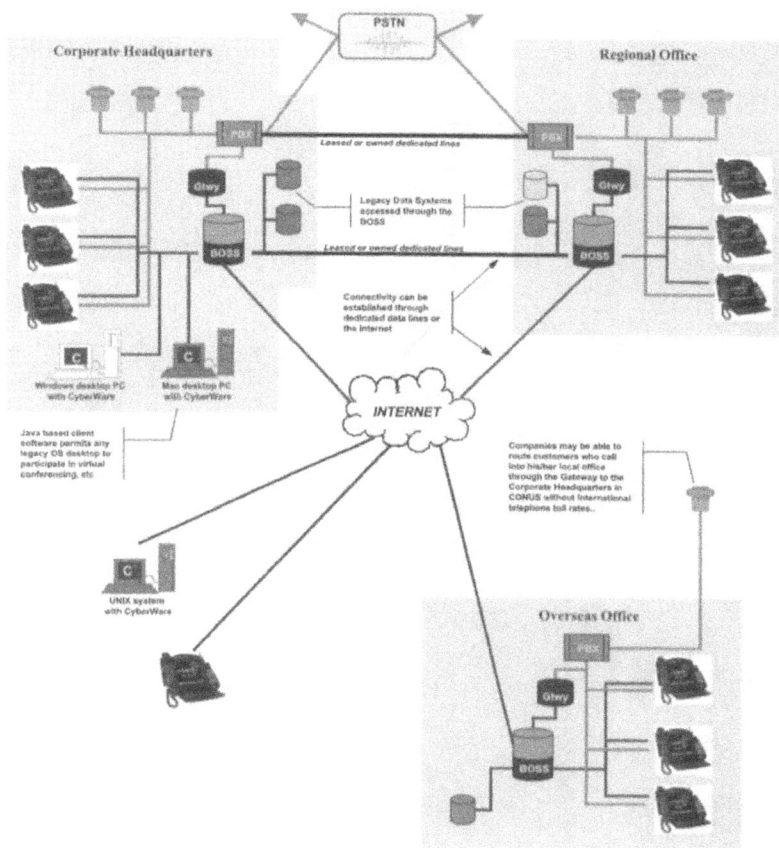

Appendix IV

The CyberFone Family of Software

CyberNet

Suite of Products

CyberNet™ Group, Inc.'s product line is directed toward applying convergent technology simply, reliably and inexpensively to the new demands of business, government, education, and the consumer. This product line consists of software, hardware, and network system overlays dedicated to virtual and real communications — people-to-people — anywhere, anytime, and is platform independent.

Virtual Meeting System™

The CyberNet Virtual Meeting System™ is the only unifying communication solution that brings people together without restrictions on their operating platform. Participants may use Windows, Macintosh, Solaris or a variety of other operating systems. The Virtual Meeting System uses the software and technology created for the CyberNet™ Communications System (CCS) and packages it in an easy-to-use, desktop computer-based software application that includes audio and video teleconferencing using Voice-over-IP (VOIP). The interactive "whiteboard" utility also enables users to share and collaboratively edit text (ASCII/RTF) and images (JPEG/GIF). A beta version of the software will be available in July 1999 with a full function version (including video) to follow in later upgrades.

CyberBOSS™

CyberBOSS™ (Broad Operations Systems Server) is the control software that connects each source to all destinations. It can be part of an ISP system, on a CyberNet™ server, or any private server or server application. CyberBOSS™ also acts as the MiddleWare layer for managing CyberFone™ client connections, interpretation of results and initial routing of data kernels. CyberBOSS™ embraces an open database philosophy, and multiple database and legacy servers can be simultaneously supported, even for different fields of the same transaction data packet. CyberBOSS™ can function on diverse server platforms.

CyberAGE™

CyberAGE™ (Application Generator Engine) is a thin client. It was developed for the CyberFone™ but can function on any client or server platform. Utilizing locally stored or downloadable application scripts, the CyberAGE™ client controls the user interface (display), manages transaction generation, initial data validation, kernel storage or transmission, and any external application or modules specified by the service partner. CyberAGE™ can function on diverse client and server platforms.

CyberNet

CyberFone™

The CyberFone™ is CyberNet™'s screen phone that incorporate the thin client CyberAGE™ together with the software that enables web access, e-mail, word processing, transactions, and linkage to the control server directed by CyberBOSS™. CyberFone™ is the functional equivalent of a network computer that incorporates full telephony and videoconferencing capabilities. The CyberFone™ unit consists of a 233MHz processor with 32Mb of RAM (expandable to 64Mb) and 12Mb of flash memory (expandable to 144Mb) for local storage. The unit's 16-bit color display measures 8" diagonally, with a screen resolution of 640 by 480 pixels. This 12" by 6" unit operates under CyberAGE™ and allows for telephone calls over the Internet, Virtual Private Networks (VPN) or over regular telephone (POTS) lines. The system includes features such as call return, call waiting, one-touch redial, programmable number storage, and optional Smart Card and swipe card capabilities.

CyberNet™ Pad

CyberNet™ has developed a Java-based word processor exclusively for use with the CyberNet™ Communication System (CCS). The CyberNet™ Pad software will be standard on all CCS machines and will allow users to seamlessly create, store, and access written documents that are stored on either CyberBOSS™ or a local file system. The possibility of document sharing and transfer is also available to users of the CCS due to the fact that CyberNet™ Pad is totally written in Java, allowing for cross-platform accessibility. This will allow users of the CCS to perform word processing functions in numerous computer environments including Solaris, Linux, Unix, Macintosh, and Windows, while being able to store and access all pertinent documents on their local CyberBOSS™ or local file system.

CyberNet™ PC Kits

CyberNet™ offers an off-the-shelf PC kit that will allow personal computer users to convert their machines to a CyberNet™ Communication System. These kits contain a copy of CyberNet™'s patented CyberAGE™ software and a video software package with a camera that will allow users access to CyberBOSS™ where they will be able to use the CyberNet™ Virtual Meeting System. The PC kits offer cross-platform capability as well, due to the Java language programming, which allows seamless conformity between the kits and operating systems like Solaris, Windows, Linux, Unix, and Macintosh.

CyberNet

CyberNet's Virtual Meeting System™

The Virtual Meeting System™ is CyberNet™ Group's answer to many of today's businesses, organizations, and consumers who want the advantages of virtual conferencing solutions, but need the ability to connect across a variety of computer platforms using such operating systems as Windows, Macintosh and Solaris.

The Virtual Meeting System™ is a software application system directed to virtual meetings associated with any document, graphics, X-Ray, scan, or diagram. Voice and video interaction are added for full communication capability.

The Virtual Meeting System™ functions on the CyberFone™ and on any PC, Macintosh, network computer, workstation, or screen phone. It can function with or without the CyberFone™ unit. By itself, the "whiteboard" portion allows users to simultaneously share and edit images and text in a paint program-type environment. Co-workers in separate offices can design business presentations together from different locations. The system also includes voice and video conferencing, text chat and file transfer.

The Virtual Meeting System™ may be used via the CyberFone™ unit or through a personal computer. In addition, users do not have to be operating on the same system. For example, one participant may be operating from a Macintosh machine, while another is using a machine with Solaris or Windows operating system. The Virtual Meeting System™ allows both users to operate regardless of operating system.

The Virtual Meeting System™ uses software and technology created for the CyberFone™ Communications System (CCS) and packages it in an easy-to-use desktop computer-based software application. This software package, available summer 1999, will include audio and video teleconferencing using Voice-over-IP (VOIP) as well as an interactive "whiteboard" utility that enables users to share and collaboratively edit text and images. A beta version of the software will be available in July 1999 with a full function version (including video) to follow in later upgrades.

The strength of the Virtual Meeting System™ is its foundation in Java software, allowing it to run on any desktop PC operating system. The interactive whiteboard allows cooperative text (ASCII/RTF) and image (JPEG/GIF) editing, which can then be saved locally by each conference participant. Each conference is controlled by a host/moderator who grants admission and passes pen control to the participants as needed.

165

CyberNet

We at CyberNet™ see the Virtual Meeting System™ as the only unifying solution that can bring people together in a virtual meeting without restriction on their operating platform. In addition to the all the benefits most virtual-conferencing software offers, you can always be sure that with the Virtual Meeting System™ you can connect with your co-workers, clients, and customers.

Requirements:

- Software:
  - The Virtual Meeting System™ software from CyberNet™ and Java VM
- Hardware:
  - Any personal computer (Intel, Macintosh, Sparc, etc.) with 32 MB RAM.
  - Microphone and speakers required for audio conferencing
  - Camera required to send video images (when video functionality is incorporated)
- Connectivity:
  - To conduct a virtual conference on the Virtual Meeting System™ all participants must connect to a common server running the CyberBOSS™ (Level I: the Moderator at a minimum)

Development:

- Whiteboard demo at Java One in San Francisco June 15-18, 1999
- Version 1 beta available August 1999
- Version 1.2 with video available late 1999

CyberBOSS™

CyberBOSS™ (Broad Operations Systems Server) serves as the network systems co-ordinator. It is the control software that connects each and every each source to all destinations, where the destinations can be other sources, or the web, or any type of application or legacy system. The source unit can be a CyberFone™, or any laptop, Macintosh, work station, PC, or network computers, all under the control of the thin client CyberAGE™. The final destination, of course, may very well be any Internet Services, legacy systems, data warehouse systems, and any desktop — and vice versa.

CyberBOSS™ can be part of businesses, schools, healthcare facilities, government agencies, and the military.

CyberBOSS™ (Broad Operations Systems Server) functions as the middleware layer between the CyberNet™ system and computer networks, enabling connectivity to back-end business and routing systems.

CyberBOSS™ is platform independent, Y2K compliant and UNICODE capable. Its open architecture allows user applications to be run on various platforms, such as Macintosh, Solaris, UNIX, Linux or Windows.

As the middleware layer between CyberFone™ and CyberAGE™, CyberBOSS™ allows access to various applications and third-tier services, such as back-end databases, servers and legacy systems.

The CyberBOSS™ architecture is a highly structured system of services, which can be added and removed, either locally or remotely. Developed services on CyberBOSS™ are modules that perform specific functions, such as Data Retrieval, User Management, Inter-Application Messaging, Error Logging, Spell Checking, as well as others. CyberBOSS™ services can also be written by CyberNet™ clients, with the use of the CyberBOSS™ API. This allows access to customer specific networks, systems and data.

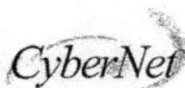

CyberNet

CyberAGE™

CyberAGE™ (Application Generator Engine) is the heart of the CyberFone™ unit. It allows the device to generate and display various on-screen templates (forms or formats). All of these can be transmitted in real time from the CyberBOSS™. This enables the user, in their own vernacular language, to select from menu choices and enter information to create the "transaction" or data base information. CyberAGE™ also serves as the vehicle for sending data back and forth between the CyberFone™ and the CyberBOSS™. The unique features of this software give the end-user the capability to use a single program to direct any and all applications.

CyberAGE™ is the presentation layer for the CyberNet™ suite of products. CyberAGE™ is built using thin client model architecture and leverages the services of the CyberBOSS™ middleware. CyberAGE™ includes web browsing, e-mail, and Personal Information Manager capabilities as standard applications.

Applications can be written for, and added to, the CyberAGE™ front-end simply and quickly by using the CyberAGE™ API supplied by CyberNet™.

All of CyberNet™'s applications are written in the Java programming language. By embracing Java, it allows CyberNet™ and their clients to draw from a large pool of proficient Java developers. This allows for easy adaptation of applications and technologies, already written in Java, which can be incorporated into the CyberAGE™ and CyberBOSS™ equation.

The Banking software application within CyberAGE™ allows for direct funds transfer, bill payment and all personal banking needs via connections to various legacy and server systems. Utilizing this Banking application, users can securely perform e-Commerce and store transactions. CyberAGE™ also allows data to move through the CyberFone™, and like the Virtual Meeting System™, it can be run through a CyberNet™ enabled personal computer as well.

CyberAGE™ is platform independent, Y2K compliant and UNICODE capable. Its open architecture allows user applications to be run on various platforms, such as Macintosh, Solaris, UNIX, Linux or Windows.

CyberFone™

The *CyberFone*™ is CyberNet™'s screen phone that incorporates the thin client CyberAGE™ together with the software that enables web access, e-mail, word processing, transactions, and linkage to the control server directed by CyberBOSS™.

The CyberFone™ unit is a 12" by 6" appliance, similar to a standard telephone, but includes the ability to conduct video conferencing, Voice over IP, document and graphic editing, Internet and e-mail capabilities, and much more. Each unit is comprised of regular telephone communication capabilities, a transaction assembly server operating system, an enhanced screen, a keyboard, and a card swipe reader.

CyberFone™ combines full computer functionality with complete telephony capability, including caller ID, call waiting and call forwarding.

Setting up CyberFone™ is as simple as plugging the unit into one standard electrical outlet and one regular POTS telephone line. It may be used independently, in tandem with a personal computer, or as part of a computer network. This combination of simplicity with familiarity will allow the user to immediately access the CyberFone™ without training.

CyberFone™ is the first communications convergence appliance for personal use within the business or home. It has the power of a desktop computer, plus the added advantage of a convenient size. Its simplicity and familiarity with 2-way video and telephony features allows the CyberFone™ to be used quickly and easily without extensive training.

This simple system enables the user to execute transactions, transmit them to a server or host, and receive data related to the transaction. An almost unlimited number of applications can be processed to multiple databases, independent of any particular operating systems.

The CyberFone™ core consists of a Cyrix MediaGX processor running at 233 MHz with 32 MB RAM. The unit's integrated display measures 8" diagonally, with a screen resolution of 640x480 pixels with 16-bit colors. These units use a laptop-sized full function keyboard and a touch screen with a stylus pen for data entry and system navigation.

CyberNet

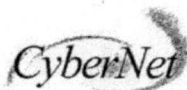

Application transaction data, whether generated locally on the CyberFone™ or on-line, will be routed to the CyberBOSS™. The CyberBOSS™ is the middleware layer that enables connectivity to back-end legacy systems and applications. By running CyberBOSS™ on your existing servers, users can be assured that any transaction data is authentic and error-free.

The world of the future will consist of transactions, which originate at the CyberFone™, and are transmitted to a host via a public-switched, cellular, Internet, Intranet, or any other form of communication network. The CyberNet™ Communications System has been developed to be the entry point with cutting-edge, cross-platform technology that allows users the ability to access information immediately.

Additionally, the CyberFone™ appliance has the advantages of simplicity and ease of use, by minimizing the need of programs continually being added to the user station and reducing the time consuming download of programs.

The general applications of the CyberFone™ unit include business, banking, medical, education, hospitality and household. Business applications include video conferencing, e-mail, web browsing and contact management. Banking applications include funds transfer, bill payment, account balance information, and loan processing. Medical applications include medical monitoring, prescription drug information and interaction, and medical records access and update. Education applications include connecting homebound students, students in remote venues and classroom students in a "Virtual Classroom." Hospitality applications include the full range of communications desired by business and leisure travelers. Household functions include online banking and shopping, weather reports, stock quotes, movie listings, school closing information and control of household devices (including security systems).

CyberNet

CyberFone™ Specifications

Description	Feature/Specification
Host processor/CPU	Cyrix Embedded 233 MHz
Memory/host	32 MB DRAM (expandable to 128 MB)
Memory/Expansion	12 MB flash memory (expandable to 16 MB), 2 MB EPRON
Display screen (LCD)	Standard - 8" diagonal DSTN VGA, 640x480, 16-bit Colors Optional - 8" diagonal TFT VGA, 800x600, 16-bit Colors
System Navigation	Stylus pen and touch panel (Optional Mouse – PS/2 or Serial)
Keyboard	Full QWERTY keyboard (English), O/S enabled, standard notebook keyboard size
Modem	56 Kbps V.90
POTS Line	One-line operation Full duplex speaker phone
I/O port	Parallel port Serial port (RS-232) USB port VGA port (external) Phone Line port (RJ-11) LAN port (RJ-45) Microphone-In port Headset/Speaker-Out port PS/2 port Infrared port
Power Module	External AC power adapter, 90~260VAC, 50/60 Hz DC to DC converter for system, LCD and battery Power consumption under normal operation less than 25W Backup battery (calculator size)
Telephone Dialer	18 Traditional telephone function keys Dialing keys (0,1,2,3,4,5,6,7,8,9,*,#) Redial key Speaker key Mute key Hold key Volume key (toggle) Flash key Serve as an input device for the address book
LED	Power Speaker-on On hold Message indicator
Soft keys	Six CyberFone programmable function keys
Integrated Camera	Tilt up, down, left and right (swivel mount)
Low power standby mode	With basic voice call available while power is stopped

CyberBOSS™ will support printers for those customers that want to use stand-alone printers.

CyberNet

FREQUENTLY ASKED QUESTIONS

How did the CyberNet™ Group, Inc. begin?

CyberNet™ began in 1995 as the vision of Dr. Rocco L. Martino, one of the world's leading experts in the creation of computer and communication systems for many of the world's largest and most complex organizations in business, government and industry. Dr. Martino, one of the original computer pioneers, believed that it is possible to create an integrated system of communications capabilities that is simple, reliable, and affordable.

Three years after establishing the patents, designing the products, and developing the software with his son Joseph Martino, Dr. Martino incorporated CyberNet™ and brought on a seasoned management team led by Navy Rear Admiral (Ret.) Robert Ellis. These men shared Dr. Martino's vision, and they became part of a close-knit group that has transformed CyberNet™ from a dream into a reality.

CyberNet™ is dedicated to empowering the business or individual customer with the most cutting-edge communications available today. With Dr. Martino serving as Chairman and CEO, RADM. Ellis as Chief Operating Officer, and Joseph Martino as Chief Technology Officer, and supported by a distinguished board of business and professional leaders, CyberNet™ is poised to lead the next generation of voice, video and data technology.

What is the CyberNet™ Communication System?

The CyberNet™ Communications System is a complete communications convergence solution that brings together all the modern capability of computers, communications and visualization. The system is a convergent technology solution leading to a virtual society, providing information on a real-time basis in a simple, inexpensive, and familiar form. The end result is 2-way interactive Video, full telephony (including Voice over IP), Internet Access, and data transactions into one package with a single terminal device. The system consists of three levels of products:

- Networks that coordinate multiple users on diverse platforms amongst themselves, and also linked to legacy systems — anywhere, anyplace, anytime, on any platform.

CyberNet

- Software packages and systems that enable the network and also the end-user client system, whether a CyberFone™, a PC, Macintosh, Network Computer, other Screen Phones, or handheld device — all of which are directed as a client by the patented CyberNet™ thin client software system CyberAGE™.
- The CyberFone™ incorporates all of the capability for end-user client applications without the need for a complex operating system, local storage, virus protection, etc. These are all reserved in a separate co-ordination server unit controlled by the CyberNet™ software system — CyberBOSS™.

All three are independent and may be used by any application — Macintosh, Solaris, UNIX, Linux, or Windows. All CyberNet™ software can be installed for operation on the CyberFone™ or on independent hardware platforms.

What is the Virtual Meeting System™?

The Virtual Meeting System™ is a software application system directed to virtual meetings associated with any document, graphics, X-Ray, scan, or diagram. Voice and video interaction are added for full communication capability.

The Virtual Meeting System™ functions on the CyberFone™ and on any PC, Macintosh, network computer, workstation, or screen phone. It can function with or without the CyberFone™ unit. By itself, the "whiteboard" portion allows users to simultaneously share and edit images and text in a paint program-type environment. Co-workers in separate offices can design business presentations together from different locations. The system also includes voice and video conferencing, text chat and file transfer.

The Virtual Meeting System™ may be used via the CyberFone™ unit or through a personal computer. In addition, users do not have to be operating on the same system. For example, one participant may be operating from a Macintosh machine, while another is using a machine with Solaris or Windows operating system. The Virtual Meeting System™ allows both users to operate regardless of operating system.

What are CyberAGE™ and CyberBOSS™?

CyberAGE™ (Application Generator Engine) is the heart of the CyberFone™ unit. It allows the device to generate and display various on-screen templates (forms or formats). All of these can be transmitted in real time from the CyberBOSS™. This enables the user, in their own vernacular language, to select from menu choices and enter information to create the "transaction" or data base information. CyberAGE™ also serves as the vehicle for sending data back and forth between the CyberFone™ and the CyberBOSS™. The unique features of this software give the end-user the capability to use a single program to direct any and all applications.

CyberNet

CyberBOSS™ (Broad Operations Systems Server) serves as the network systems co-ordinator. It is the control software that connects each and every source to all destinations, where the destinations can be other sources, or the web, or any type of application or legacy system. The source unit can be a CyberFone™, or any laptop, Apple, work station, PC, or network computers, all under the control of the thin client CyberAGE™. The final destination, of course, may very well be any Internet Services, legacy systems, data warehouse systems, and any desktop — and vice versa.

CyberBOSS™ can be part of businesses, schools, healthcare facilities, government agencies, and the military.

What is CyberFone™?

The CyberFone™ is CyberNet™'s screen phone that incorporate the thin client CyberAGE together with the software that enables web access, e-mail, word processing, transactions, and linkage to the control server directed by CyberBOSS™.

The CyberFone unit is a 12" by 6" appliance, similar to a standard telephone, but includes the ability to conduct video conferencing, Voice over IP, document and graphic editing, Internet and e-mail capabilities, and much more. Each unit is comprised of regular telephone communication capabilities, a transaction assembly server operating system, an enhanced screen, a keyboard, and a card swipe reader.

CyberFone™ combines full computer functionality with complete telephony capability, including caller ID, call waiting and call forwarding.

Setting up CyberFone™ is as simple as plugging the unit into one standard electrical outlet and one regular POTS telephone line. It may be used independently, in tandem with a personal computer, or as part of a computer network. This combination of simplicity with familiarity will allow the user to immediately access the CyberFone™ without training.

What are the advantages of the CyberNet™ Communication System vs. a PC and a phone?

Cyberfone™'s major advantages over the personal computer are simplicity, ease of use, obsolescence reduction, and significant cost reduction over the life of the system. Set-up requires only a standard phone hook-up and A/C outlet for now. However, later versions will include cellular and cordless systems.

To operate CyberFone™, simply turn the phone on and log in. The executive menu screen immediately appears, and this directs the user to the Internet, e-mail, word processing, virtual meetings, or to applications. The CyberAGE™ thin client will handle all aspects of the routing and interaction between user and the system.

The system is always ready for use. There are no complicated instruction manuals, tangled wires, and intricate installation directions.

CyberNet

CyberFone™'s size, which is closer to a telephone than a personal computer, makes it ideal for both businesses and households. CyberFone™ is also substantially less expense, not only in its price tag, but in its maintenance costs as well. Connection to CyberBOSS™ allows users to download programs from the server as they become available instead of having to purchase costly upgrades, as is the case with most personal computers.

Other advantages include: remote storage from creation, integrated communications via multiple paths, state of the art security, and point of use system.

No more backups, no more disc crashes, no more virus problems at the user location — and the ability to communicate anywhere, anytime, with anyone.

For transactions through CyberFone™, is there an extra charge for use of remote server?

The CCS is a thin client/server system and features a Java-based solution. Its software remains platform independent and permits access to any legacy system without reprogramming or expensive retrofits.

This allows CyberFone™ to operate as part of an overall system — whether banking, healthcare, education or others — that functions in combination with remote servers and one or more CyberFone™ units.

The system, as such, is essentially the province of the provider of that particular service. Depending on the policies of the provider, transactions using the remote server may or may not carry a fee. CyberFone™ is independent of these considerations and can operate whether there is a fee or not.

CyberNet

Are the CyberNet™ products ready for use?

The Virtual Meeting System™, video conferencing, medical, educational and banking applications are all available now for demonstrations and installation. This can be direct installation, or via portals utilizing equipment in our Pennsylvania office and/or on remote site under our control.

The CyberNet™ software systems built around CyberBOSS™ and CyberAGE™ are available for installation and/or license now.

The CyberFone™ unit is available for test now and for installation in mid-summer. Currently we are under contract with the Scientific Application International Corporation (SAIC) for a fall installation of CyberFone™ and CyberBOSS™ under the Hospitals, Universities, Businesses, and Schools (HUBS) program.

What are some of CyberNet™'s applications?

CyberNet™ is designed to be the communications system for the next millennium. It can either augment your current telephone and/or personal computer systems, or CyberNet™ products can operate independently as fully integrated separate units. The options for the CyberNet™ family of products are virtually endless.

Some of CyberNet™'s product application areas include:

- Business

  Conduct video conferences with clients and employees anywhere
  Conduct cheaper telephone conference calls via Voice over IP
  Allows multiple parties to edit and work on the same document or graphic from different machines in remote locations
  Internet and e-mail capabilities

- Education

  Allows teachers to conduct lectures with students in remote locations
  Allows students to participate and receive assistance with homework from their computers

- Healthcare

  Access to medical records
  Facilitate the check-in process at the doctor's office
  Consult your physician from home
  Allows doctors in remote locations to consult one another

- Hospitality

  Make hotel/airline reservations and confirmations
  Access e-mail on the road
  Edit business presentations

CyberNet

- Personal Banking
  - Funds transfer
  - Bill payment
  - Account balance information
  - Loan processing

- Government/Military
  - Government-to-government connectivity at all levels

- E-Commerce
  - Make on-line purchases easily and securely

- Consumer
  - Get stock quotes, movie listings, weather reports, and news
  - Communicate with faraway friends and family visually
  - Utility meter monitoring

Appendix V

Diagrams Illustrating the UNI-LINK®
Virtual Application Platform

CYBER TECHNOLOGY GROUP INC.

CyberFone Technologies
Cyber Technology Group

Presentation Includes Diagrams Illustrating the
UNI-LINK®
Virtual Application Platform

October, 2009

CYBER TECHNOLOGY GROUP INC.

Table of Contents

181

Table of Contents (con't.)

2

CYBER TECHNOLOGY GROUP INC.

CYBER TECHNOLOGY GROUP INC.

CTG Offers Simplicity

Patented UNI-LINK technology and software solutions that enable ubiquitous and secure real-time computing.

Today:

- Secure instant messaging
- Rapid mobile enablement of existing applications

Tomorrow: *Elegant path to emerging capabilities*

- Enhanced real-time messaging that integrates voice, video, and data
- Grid computing applications
- Mobile commerce

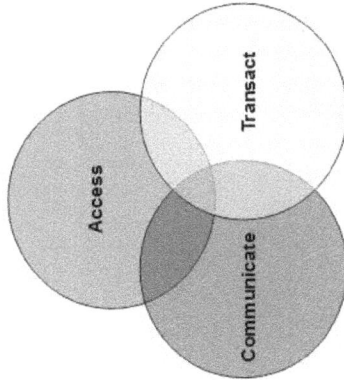

Access

Transact

Communicate

3

CYBER TECHNOLOGY GROUP INC.

UNI-LINK®
Architecture

184

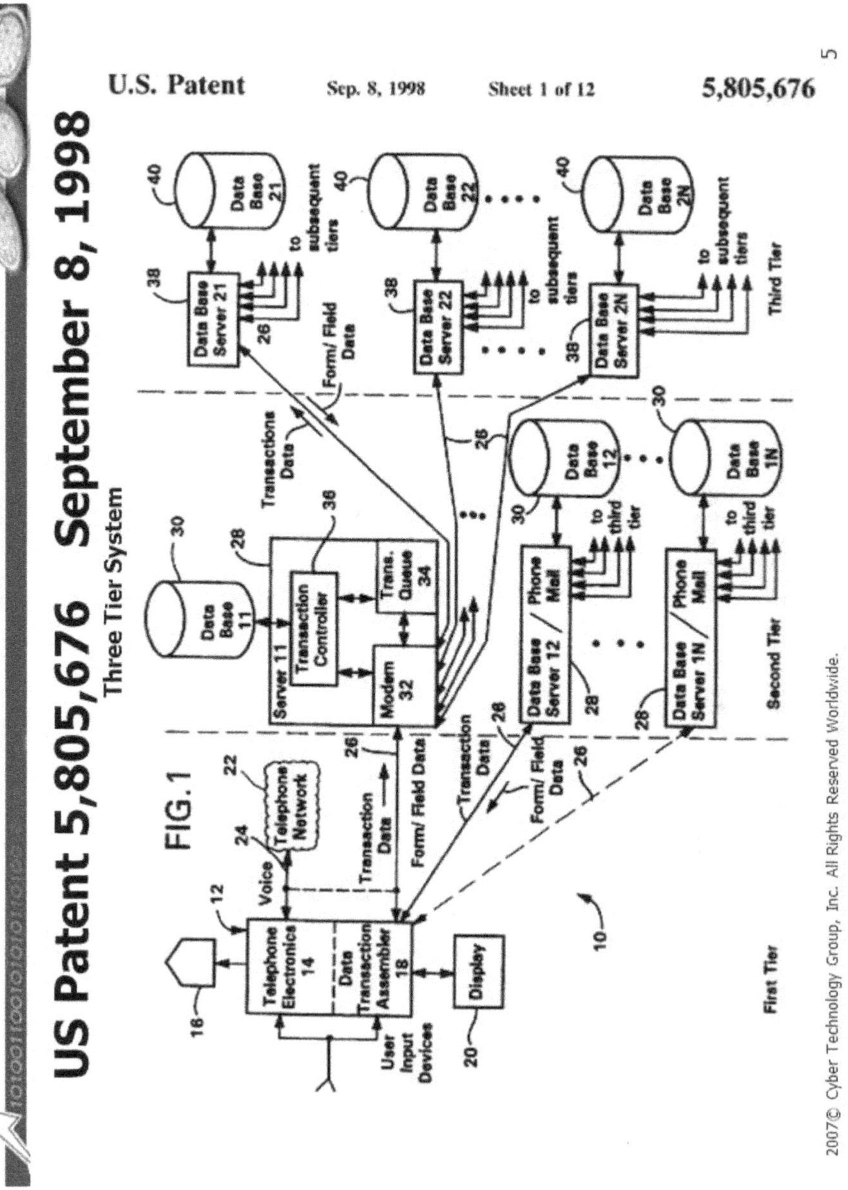
CYBER TECHNOLOGY GROUP INC.
US Patent 5,805,676 September 8, 1998
Three Tier System
U.S. Patent
Sep. 8, 1998
Sheet 1 of 12
5,805,676
FIG.1
Telephone Electronics 14
Data Transaction Assembler 18
Display
User Input Devices
Voice
Telephone Network
Transaction Data
Form/ Field Data
Server 11
Transaction Controller
Trans. Queue 34
Modem 32
Data Base 11
Transactions Data
Form/ Field Data
Data Base Server 12
Phone / Mail
Data Base 12
Data Base Server 1N
Data Base 1N
to third tier
Data Base Server 21
Data Base 21
Data Base Server 22
Data Base 22
Data Base Server 2N
Data Base 2N
to subsequent tiers
First Tier
Second Tier
Third Tier
2007© Cyber Technology Group, Inc. All Rights Reserved Worldwide.
5

ROCCO LEONARD MARTINO

CYBER TECHNOLOGY GROUP INC.

UNI-LINK® Architecture

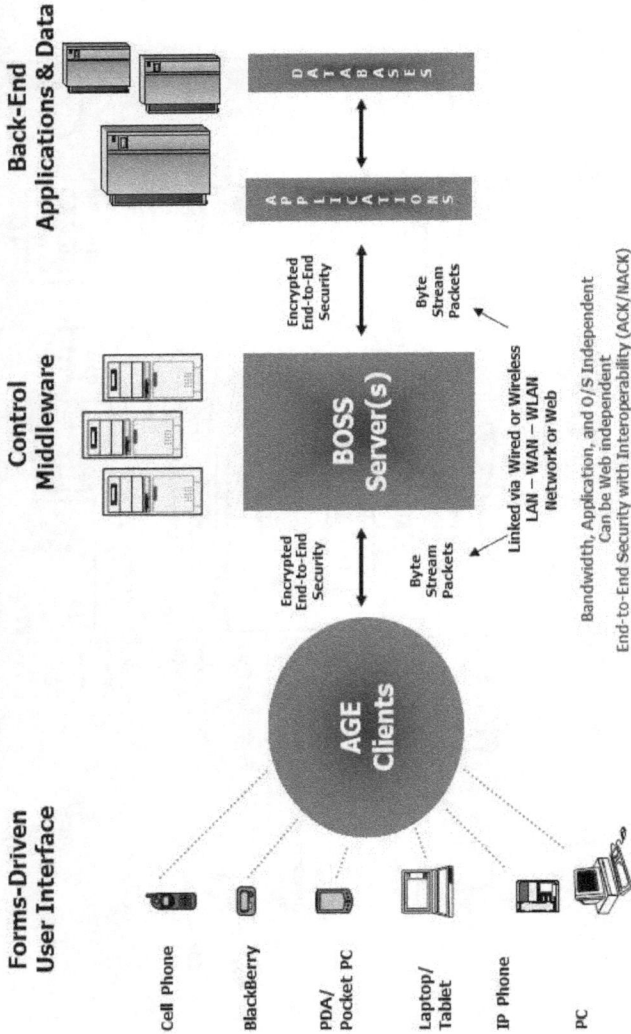

Forms-Driven User Interface

Control Middleware

Back-End Applications & Data

Cell Phone
BlackBerry
PDA/ Pocket PC
Laptop/ Tablet
IP Phone
PC

AGE Clients

BOSS Server(s)

Encrypted End-to-End Security

Byte Stream Packets

Encrypted End-to-End Security

Byte Stream Packets

Linked via Wired or Wireless LAN – WAN – WLAN Network or Web

Bandwidth, Application, and O/S Independent
Can be Web independent
End-to-End Security with Interoperability (ACK/NACK)

AGE: Application Generator Engine
BOSS: Broad Operation System Server
2007© Cyber Technology Group, Inc. All Rights Reserved Worldwide.

186

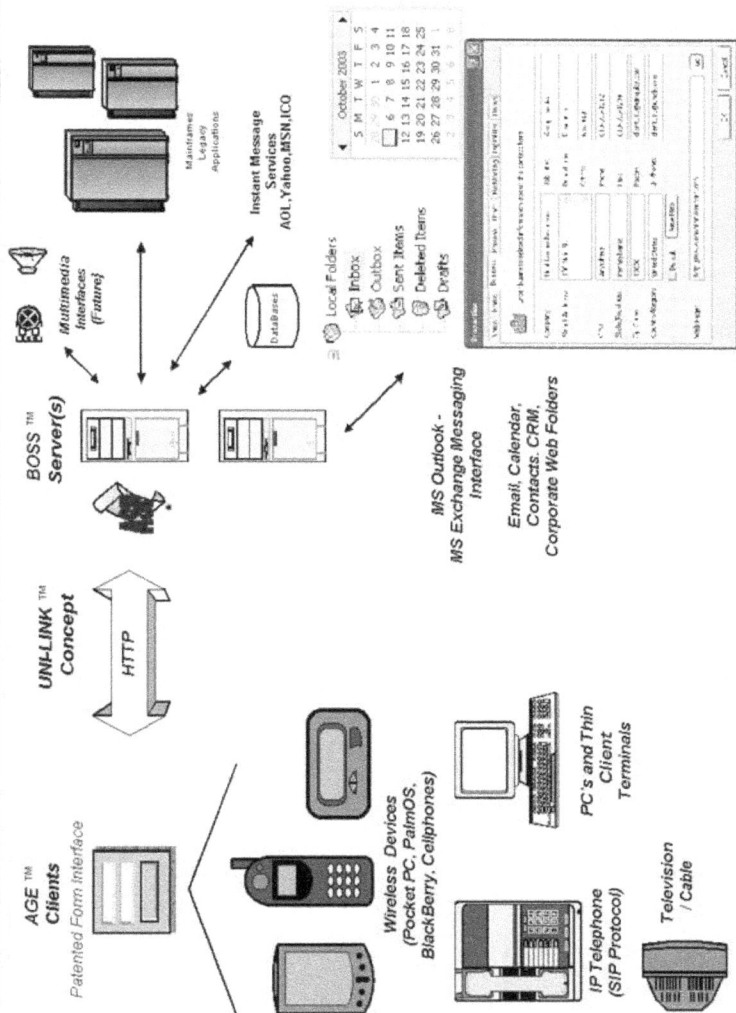

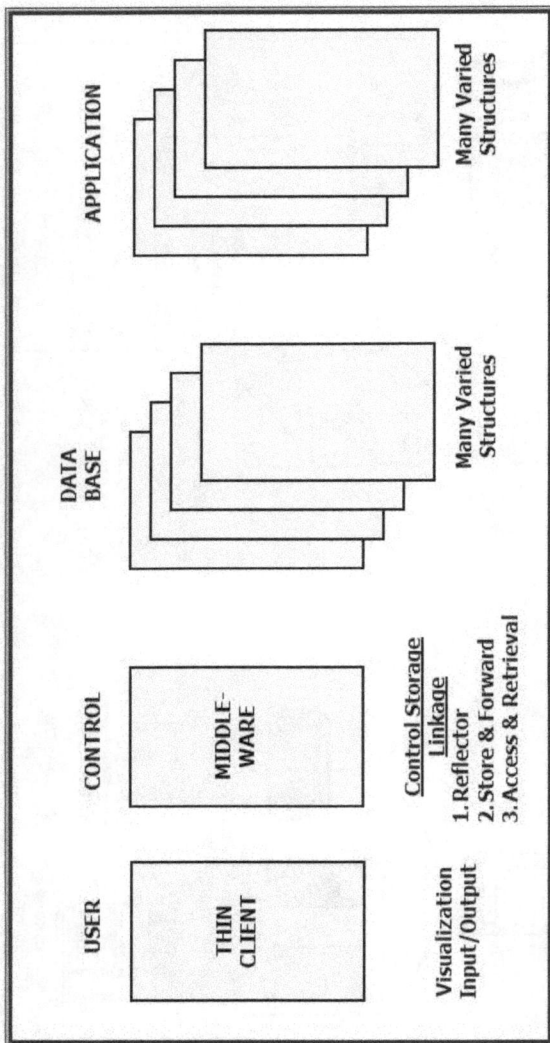

CYBER TECHNOLOGY GROUP INC.

Solutions

The ROCCI Integrated Solution

ROCCI integrates multiple disparate technologies to deliver powerful new communications capability to the point of need

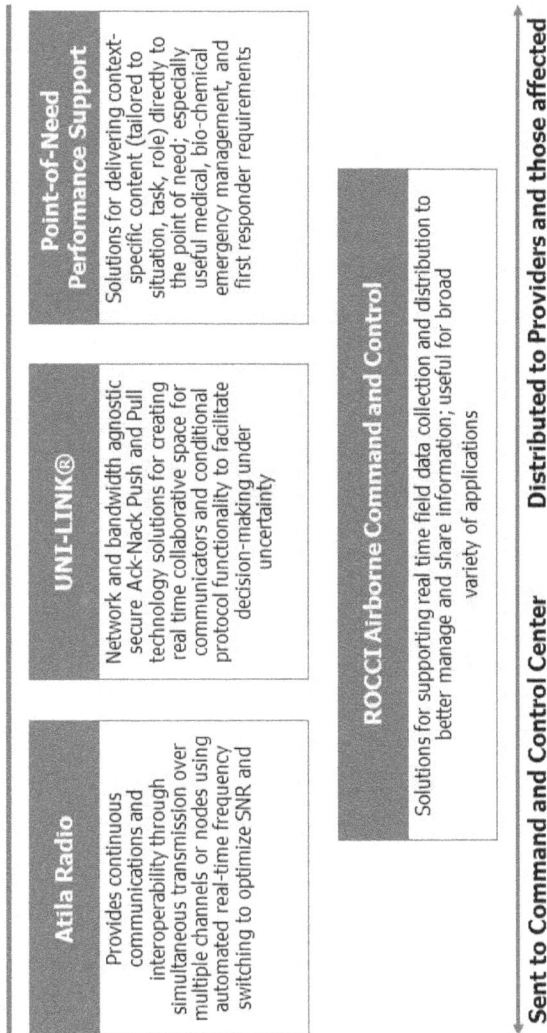

Atila Radio

Provides continuous communications and interoperability through simultaneous transmission over multiple channels or nodes using automated real-time frequency switching to optimize SNR and

UNI-LINK®

Network and bandwidth agnostic secure Ack-Nack Push and Pull technology solutions for creating real time collaborative space for communicators and conditional protocol functionality to facilitate decision-making under uncertainty

Point-of-Need Performance Support

Solutions for delivering context-specific content (tailored to situation, task, role) directly to the point of need; especially useful medical, bio-chemical emergency management, and first responder requirements

ROCCI Airborne Command and Control

Solutions for supporting real time field data collection and distribution to better manage and share information; useful for broad variety of applications

Sent to Command and Control Center   **Distributed to Providers and those affected**

9

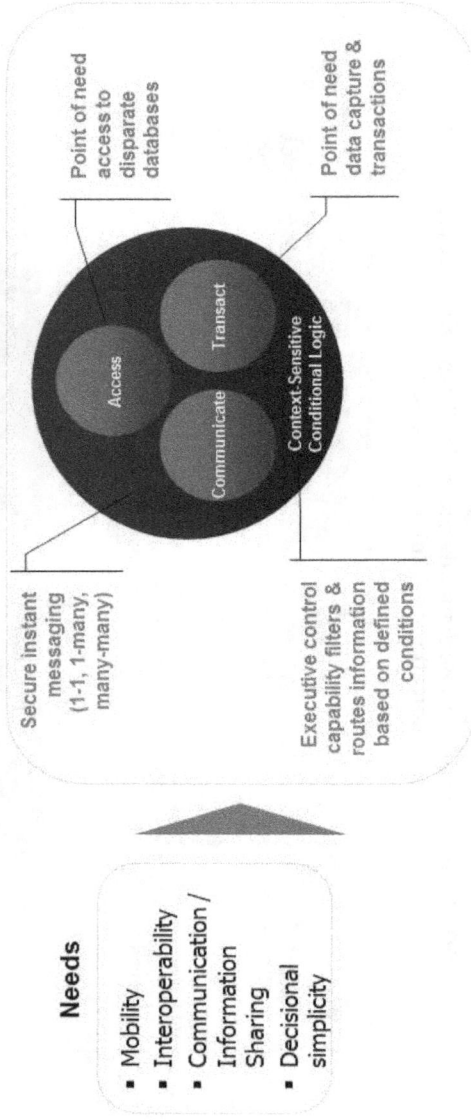

CYBER TECHNOLOGY GROUP INC.

STEVENS Institute of Technology

CyberFone Technologies' UNI-LINK® and Stevens Institute's Attila Technologies offer patented solutions for context-sensitive information delivery to the point of need.

10

CTG Core Capabilities

- Point of need access to disparate databases
- Point of need data capture & transactions
- Secure instant messaging (1-1, 1-many, many-many)
- Executive control capability filters & routes information based on defined conditions

Access
Transact
Communicate
Context-Sensitive Conditional Logic

Needs

- Mobility
- Interoperability
- Communication / Information Sharing
- Decisional simplicity

Core capabilities and network and frequency independence enhance performance at the point of need.

CYBER TECHNOLOGY GROUP INC.

UNI-LINK® System

DHS/FEMA Command and Control in Real-Time
Patented Conditional Logic-Based Instant Messaging . . .
Secure Messages and Transactions

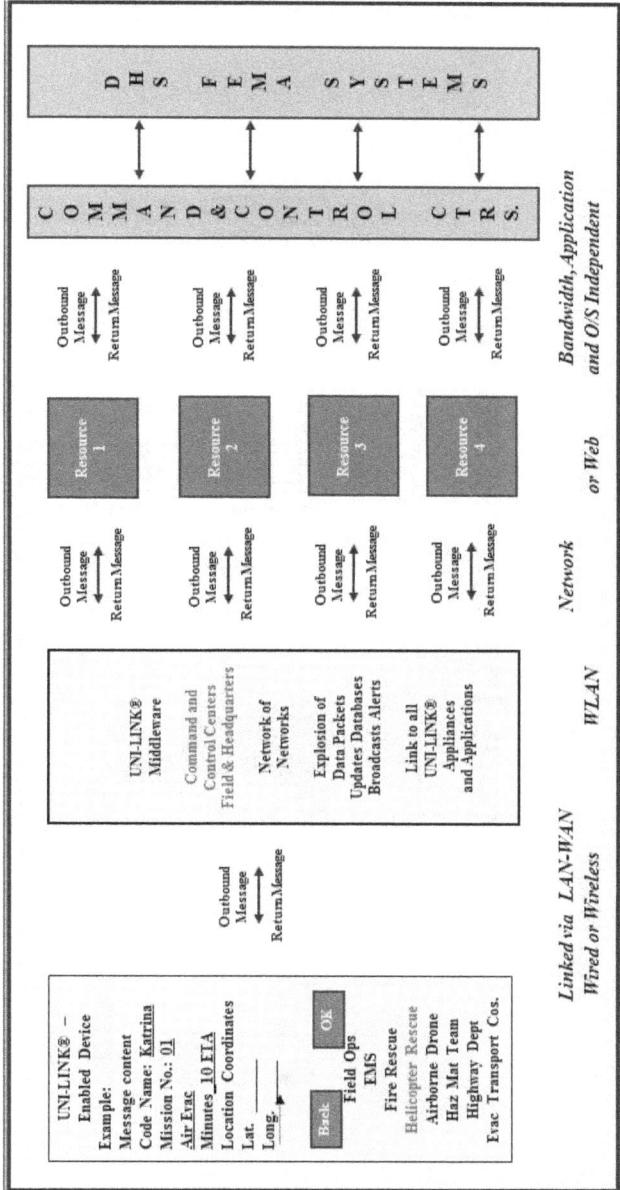

UNI-LINK® –
Enabled Device
Example:
Message content
Code Name: Katrina
Mission No.: 01
Air Evac
Minutes 10 ETA
Location Coordinates
Lat. _____
Long. _____

Back   OK

Field Ops
EMS
Fire Rescue
Helicopter Rescue
Airborne Drone
Haz Mat Team
Highway Dept
Evac Transport Cos.

UNI-LINK®
Middleware

Command and
Control Centers
Field & Headquarters

Network of
Networks

Explosion of
Data Packets
Updates Databases
Broadcasts Alerts

Link to all
UNI-LINK®
Appliances
and Applications

Outbound
Message
Return Message

Resource 1
Resource 2
Resource 3
Resource 4

Outbound Message / Return Message

COMMAND & CONTROL CTRS.

DHS FEMA SYSTEMS

*Linked via LAN-WAN
Wired or Wireless*   *WLAN*   *Network
or Web*   *Bandwidth, Application
and O/S Independent*

2007© Cyber Technology Group, Inc. All Rights Reserved Worldwide.

11

191

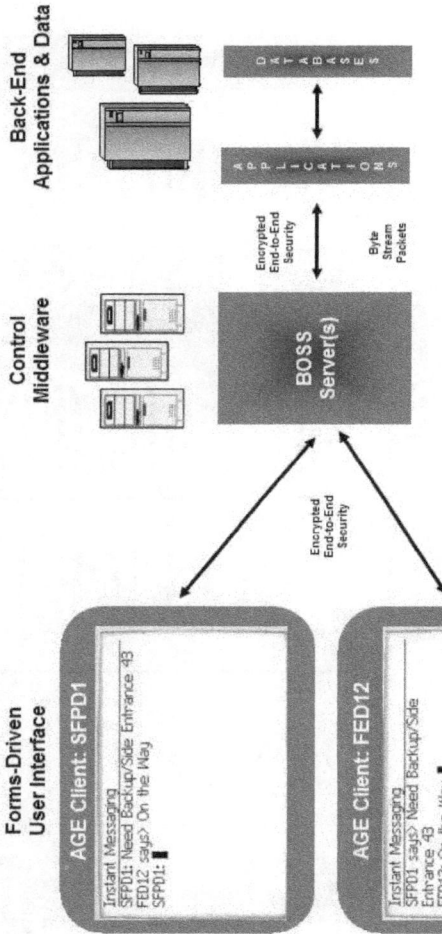

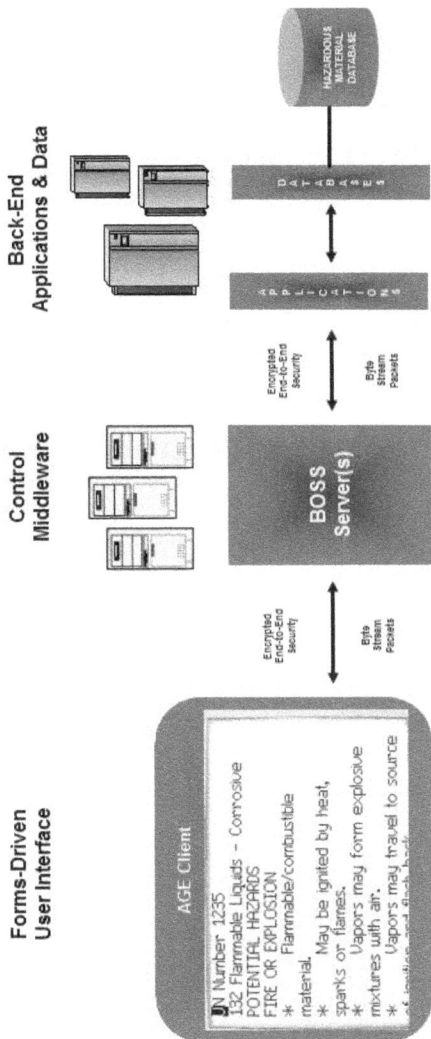

CYBER TECHNOLOGY GROUP INC.

Secure Remote Access to Diverse Data

EXAMPLE

Forms-Driven User Interface

Control Middleware

Back-End Applications & Data

AGE Client

UN Number 1235
132 Flammable Liquids - Corrosive
POTENTIAL HAZARDS
FIRE OR EXPLOSION
* Flammable/combustible
material.
* May be ignited by heat,
sparks or flames.
* Vapors may form explosive
mixtures with air.
* Vapors may travel to source

Encrypted End-to-End Security

Byte Stream Packets

BOSS Server(s)

Encrypted End-to-End Security

Byte Stream Packets

APPLICATIONS

DATABASES

HAZARDOUS MATERIAL DATABASE

13

Structured Access and Transactions with Diverse Data

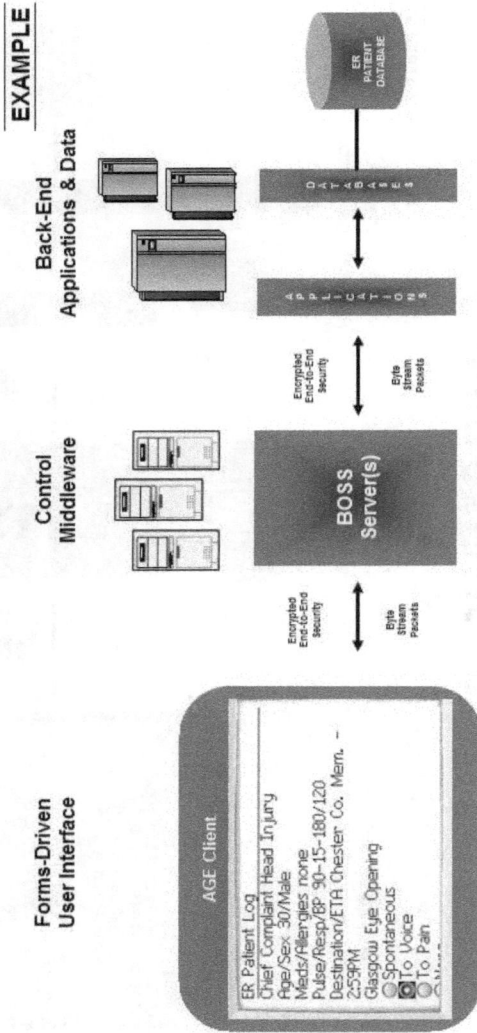

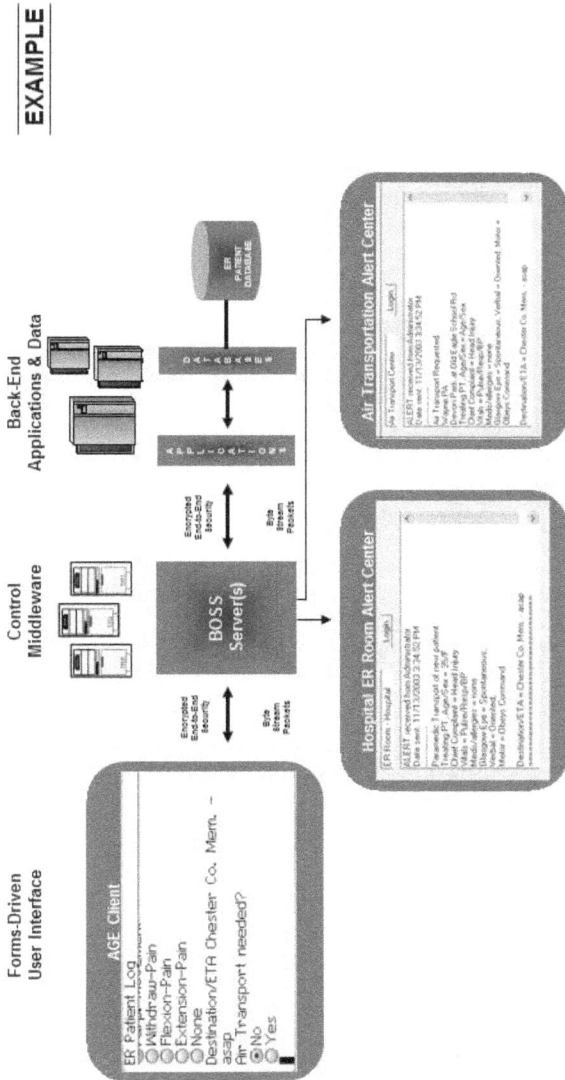

CYBER TECHNOLOGY GROUP INC.

UNI-LINK®

Forms-Driven User Interface
Ground, Sea, Air

Command & Control Centers
Field and Headquarters

Centers DOD Systems

AGE Clients

BOSS Servers (Network of Networks)

APPLICATIONS

DATABASES

Encrypted End-to-End Security

Byte Stream Packets

Encrypted End-to-End Security

Byte Stream Packets

Linked via Wired or Wireless
LAN – WAN – WLAN
Network or Web

Soldiers
Shooters
Helicopter
Airplane
Tank

Bandwidth, Application, and O/S Independent
Can Be Web Independent
End-to-End Security with Interoperability (ACK/NACK)

AGE: Application Generator Engine   BOSS: Broad Operation System Server
2007© Cyber Technology Group, Inc. All Rights Reserved Worldwide.

16

196

CYBER TECHNOLOGY GROUP INC.

UNI-LINK®

Real time, secure transaction entry device and middleware system for entering POS transaction data into databases and sending/receiving voice, data, graphic and video data.

APPLICATIONS

POS Link

Customer File

Rebate and Coupon System

Supplier Systems

J&R Store Inventory

Product Data Description

J&R Accounting

Encrypted End-to-End Security

Byte Stream Packets

UNI-LINK® Middleware

CyberBOSS Server
(Broad Operation System Server)
POS control & DB linkages

Networks with Explosion of Data Packets
Link to all J&R Kiosks, POS Applications, And Supplier Systems

Linked via Wired or Wireless
LAN – WAN – WLAN
Network or Web

Encrypted End-to-End Security

Byte Stream Packets

Retail Kiosks And POS Capability
Powered by

CYBERAGE
(Application Generator Engine)

Touch Screen Monitor

With Touch Screen
Keyboard
Card Reader
IR Bar code/RFID
Check Printer

Bandwidth, Application, and O/S Independent
Web independent – Web Inclusive
End-to-End Security with Interoperability

CYBER TECHNOLOGY GROUP INC.

UNI-LINK®

Clinical Trial Communication and Control in Real-Time Patented Instant Messaging . . .
and Beyond– *Encrypted Text, e-mail, e-fax Messages and Transactions*

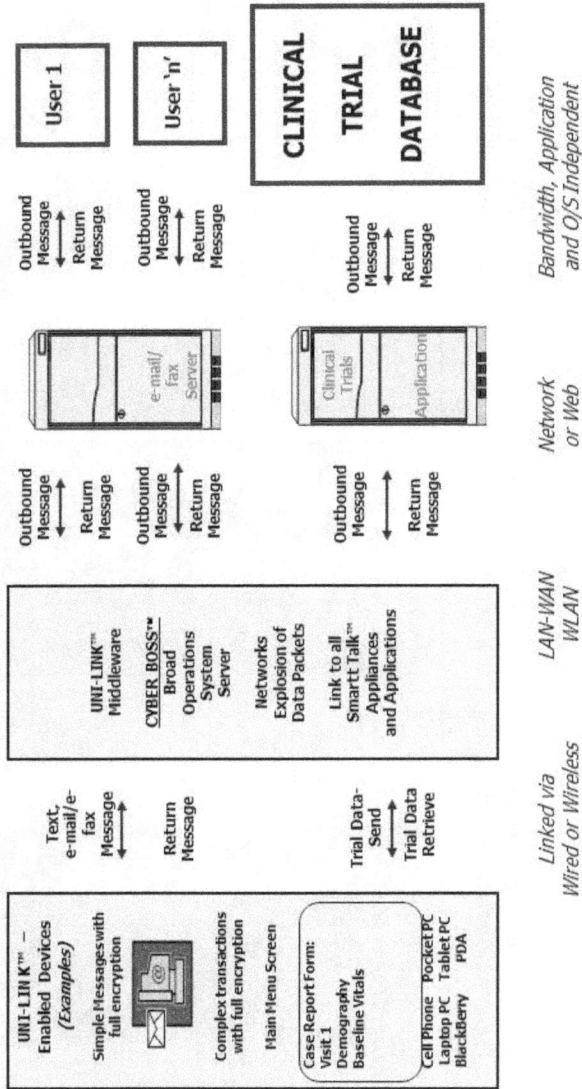

18

| | | User 1 |
| | | User 'n' |

CLINICAL TRIAL DATABASE

Outbound Message / Return Message

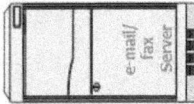

Outbound Message / Return Message

Outbound Message / Return Message

e-mail/fax Server

Clinical Trials Application

Outbound Message / Return Message

Outbound Message / Return Message

Outbound Message / Return Message

UNI-LINK™ Middleware

CYBER BOSS™ Broad Operations System Server

Networks Explosion of Data Packets

Link to all Smartt Talk™ Appliances and Applications

Text, e-mail/e-fax Message

Return Message

Trial Data-Send

Trial Data Retrieve

UNI-LINK™ – Enabled Devices *(Examples)*

Simple Messages with full encryption

Complex transactions with full encryption

Main Menu Screen

Case Report Form:
Visit 1
Demography
Baseline Vitals

Cell Phone   Pocket PC
Laptop PC   Tablet PC
BlackBerry   PDA

Bandwidth, Application and O/S Independent

Network or Web

LAN-WAN WLAN

Linked via Wired or Wireless

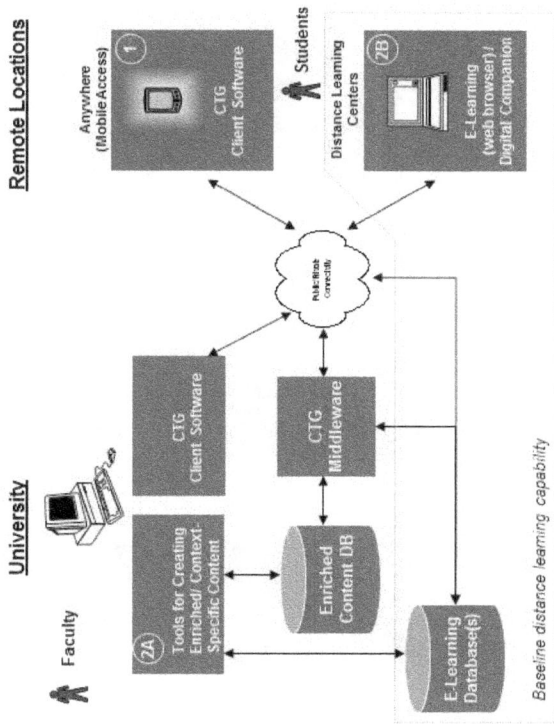

19

199

CYBER TECHNOLOGY GROUP INC.

Learning Simulations

CTG has developed an innovative approach for creating and conducting learning simulations of decision making under uncertainty.

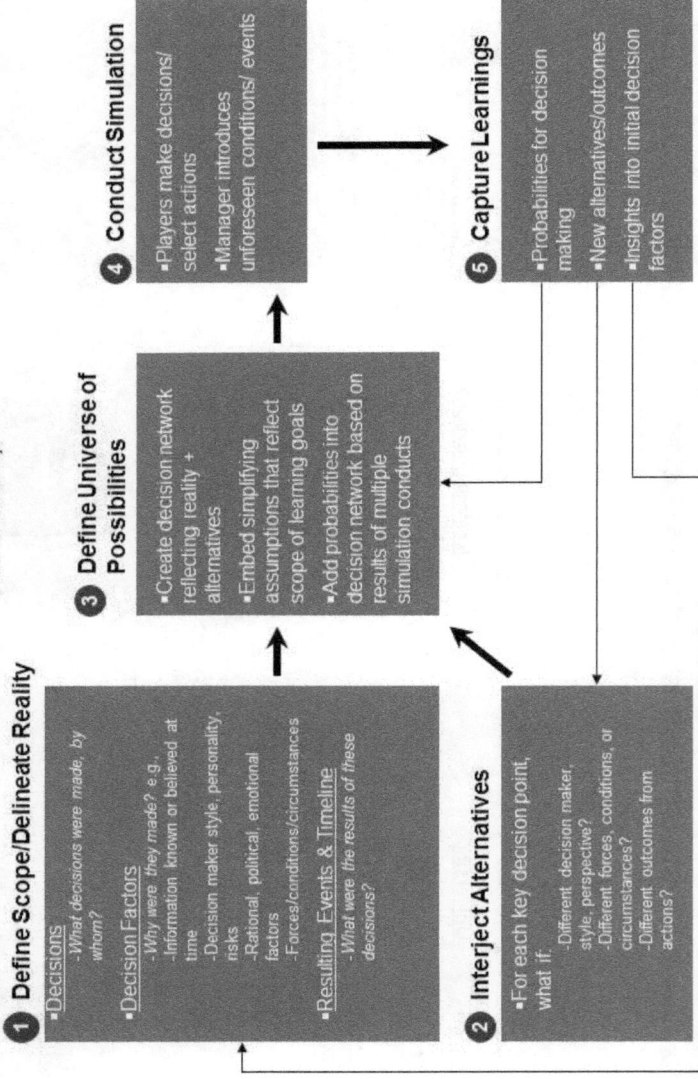

1 Define Scope/Delineate Reality

- Decisions
  - *What decisions were made, by whom?*
- Decision Factors
  - *Why were they made? e.g.,*
  - *Information known or believed at time*
  - *Decision maker style, personality, risks*
  - *Rational, political, emotional factors*
  - *Forces/conditions/circumstances*
- Resulting Events & Timeline
  - *What were the results of these decisions?*

2 Interject Alternatives

- For each key decision point, what if:
  - *Different decision maker, style, perspective?*
  - *Different forces, conditions, or circumstances?*
  - *Different outcomes from actions?*

3 Define Universe of Possibilities

- Create decision network reflecting reality + alternatives
- Embed simplifying assumptions that reflect scope of learning goals
- Add probabilities into decision network based on results of multiple simulation conducts

4 Conduct Simulation

- Players make decisions/ select actions
- Manager introduces unforeseen conditions/ events

5 Capture Learnings

- Probabilities for decision making
- New alternatives/outcomes
- Insights into initial decision factors

20

CYBER TECHNOLOGY GROUP INC.

UNI-LINK®

The Paradigm Shift to the Virtual Application

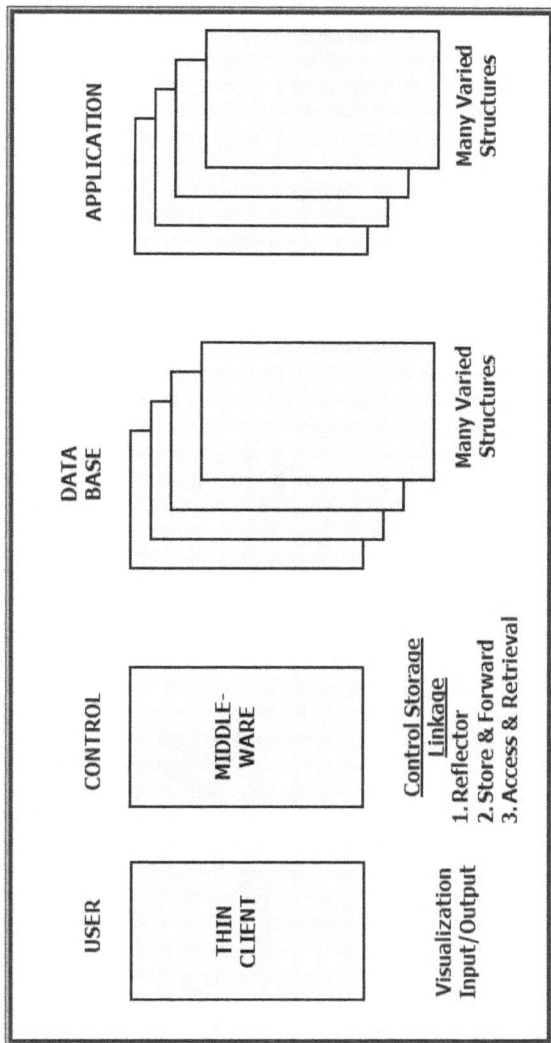

USER	CONTROL	DATA BASE	APPLICATION

THIN CLIENT

MIDDLE-WARE

Many Varied Structures

Many Varied Structures

Visualization
Input/Output

Control Storage
Linkage
1. Reflector
2. Store & Forward
3. Access & Retrieval

21

Secure Remote Access to Diverse Data

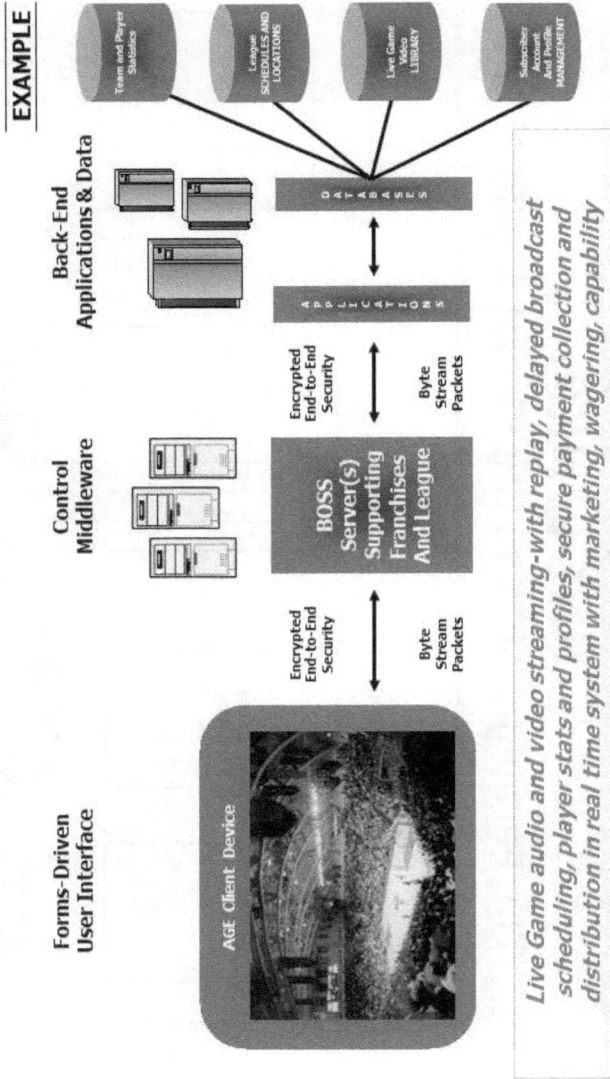

CYBER TECHNOLOGY GROUP INC.

AGE Tested Capabilities

	Laptop (Virtual AGE – http)	BlackBerry (GPRS)	Cellular Phone (with AGE)	PocketPC (WiFi)
Hazmat Lookup	Yes	Yes	Yes	Yes
ER Multiple Alerts	Yes	Yes	Yes	Yes
SMS Send	Yes	Yes	Yes	Yes
E-mail: Send	Yes	Yes	Yes	Yes
E-mail: List, Check	Yes	Yes	Yes	Yes
IM: Send	Yes	Yes	Yes	Yes
IM: Respond	Yes	Yes	Yes	Yes

23

24

Summary

UNI-LINK® Patented Technology has established a forms-based Virtual Application Platform that provides a new paradigm for all applications- simple or complex- from simple secure real time messaging to logic- based pervasive computing- wired or wireless.

The patented thin client three tier system is supplemented by an Application Creator Toolkit that enables end users without programming skills to easily and quickly create their own applications at a lower cost and quicker turnaround time than otherwise possible.

Listed below is a sampling of application types enabled by UNI-LINK®

CYBER TECHNOLOGY GROUP INC.

Sample Application Types
Wired/Wireless

General functionality
- VoIP, Set top Box Video Control, Real time secure payment systems, Audible Menus converted to Visible Menus

Emergency Management/Homeland Security
- Disaster Planning, Response, Recovery
- Maritime Port Security
- Integrate Airborne/Land-based wireless communication

Electrical Energy Consumption Efficiency Optimization
- National Grid of sensor-controlled Management and Use
- Industrial, Commercial and Residential Application of Low-cost Identical technology

Real time, Secure Transaction Processing
- Banking, Remote patient monitoring, diagnosis, prescription
  Home security system and general appliance devices control
  Lottery Wagering, Enhanced Remote Learning/Simulations
  Voting, Voting Information Management, Video On Demand
  Retail Instant Rebate and custom Marketing at the Point of Sale

Military
- Navy On Ship Communications
- Army Base and Field Operations/ Advanced Weapons Command and Control

25

205

Appendix VI

**The Martino IP Technology
An Historic Document of Capability**

The Martino IP Technology

Dr. Rocco L. Martino

CyberFone Technologies, Inc.

October 2009

Summary

The Martino Technology is directed to convergent technology, unifying computer power, universal communication capability, and visualization technology. The Martino IP covers the new wave of business expansion of two-way communication especially that centered on mobile cellular phone technology. The patents and their claims cover the system and methodology of how to provide two-way communication, wired and unwired, in the initiation and consummation of digital transactions. The patented methodology creates a virtual application platform. The invention and the granted claims cover:

1. convergent technology with any type of transaction

    (a) these two-way transactions can be transmitted with or without the form used to create it;

    (b) can include data streams representing any digital media – text, voice, video, imaging, music, pictures, or combinations of these and other media

(c) can be transmitted on wired or wireless networks

2. these transactions are created in the client device using a thin client module which is referred to in the patent applications and in the claims as a form driven operating system, or TAS – Transaction Assembler Server

(a) are independent of the operating system of the initiating device, dependent only on the use of TAS, a form driven operating system, or some thin client module that acts as outlined in the disclosure and as covered by the claims; as for example WAP which acts as a form driven operating system, and is so described in the literature and patent applications of Openwave the licensor of WAP licenses and technology

(b) TAS, or the form driven operating system, is a module that resides in the device used to originate the transaction, and which can also receive data resulting from actions initiated by the entry of data into the device that is TAS-enabled.

(c) any module whatsoever that creates a virtual application platform as does TAS infringes these patents and the claims centered on

TAS; as for example WAP, BREW, and others of the same methodology

3. The transactions are created by TAS utilizing the methodology covered by the patents and the claims

   (a) a menu is presented by TAS on the display

   (b) the menu is either generated by TAS from a data stream stored in the machine or received from an external source; or, the menu is presented in its completed form

   (c) selection of a menu option initiates another menu, a form for a transaction, or a process

   (d) if a menu is presented, it recycles back to item (c)

   (e) if a form is selected, the form is presented as in (b) and the operator answers prompts to enter data

   (f) the data entered can be a single entry (like a telephone number, URL, etc.) or multiple entries – the patents call for at least one entry, and this can be supplied by TAS in a recycling process

   (g) the data entered can identify a process to be initiated, just as a menu can accept entry of a process to be initiated; and the process can be, for example, to send the transaction, cancel it, or go to another

(h) the methodology of creating TAS or a TAS-equivalent, whether using Java, J2ME, or APIs does not absolve the infringer; i.e., the bells and whistles of a particular accomplishment do not change the dependence on the patented methodology

4. the transaction is transmitted to a control server normally remote from the device with TAS imbedded, thus creating the ability to transmit from any device to one or more devices through the control server or middleware, whether that device is TAS-enabled or not

5. the control server is linked to any number of servers and databases no matter where located in a dynamic fashion for the two-way transmission of information between any and all devices on the network whether the network is fixed or created ad-hoc

6. encryption, battery operation, telephone line use, and other capabilities are covered within the disclosure and claims. These provide a full generalization of coverage for any kind of digital transaction in any media or combination in a wired or wireless mode, based essentially in the use of a TAS-like module in the user device

linked to a control server that in turn is linked to the world.

Based upon the brief analysis of the coverage of the Martino IP technology, it is believed that infringement is widespread, and that licensing can be a significant revenue source.

This summary of capability is directed to establishing the scope of the patent and claims coverage, and to identify infringers and infringement areas; and to verify such identification by reference to the patent disclosures and the claims.

The Martino IP Technology extends the general-purpose concept of the microprocessor (and the digital computer) into unlimited applications, embodying a virtual application capability with a multi-tier approach. There is no limit to the range of applications that can be implemented with this technology, all utilizing a single thin-client server resident in the user interface. In that sense, applications are virtual since they do not require major efforts to implement, maintain, or change. The resultant applications are interoperable, wired or wireless, and operating system agnostic. That has created the explosion of the cell phone as the communications instrument of choice in a wide-variety of transactions.

The impact of this technology is analogous to the replacement of the propeller-driven aircraft with a jet. Jets have fewer moving parts, longer maintenance cycles, longevity, and lower operating costs. The same is true for systems and applications implemented with the Martino IP-protected technology. Furthermore, the functionality of the patented three tier system across a grid of computer networks in real time facilitates cloud and pervasive computing techniques regardless of application type. This patented system provides device and database independence and interoperability; and offers an integrated transaction processing system that supports:

- Cloud and Pervasive Computing Systems

- Grid Computing,

- Complex Messaging

- Purchase and supply

- Remote operations in education, medicine, merchandising

- IP Telephony,

- Conditional logic-based messaging that incorporates voice, data and video technologies.

- Directed advertising

_block

The

- The new projected wave of ultra-miniaturized devices employing Nano technology

Revenue and Profit Potential

Revenue and profit can be generated by:

1. Creating business products and services that utilize this technology under the umbrella of patent protection to forestall competition and copiers.

2. License the technology

3. Recover royalties due from those who are using the technology without paying royalties. Such revenue can be significant.

There is currently a timely opportunity to generate significant revenue and profit for new services and products without competitive pressures. The timing is quite similar to that of the early days of the computer and information industry. The Martino IP Technology can provide the impetus for the new wave of mobile multi-media communication and transaction systems.

The Martino IP technology was created to meet this potential. Patent application began on May 19th, 1995. Patents have been granted in the United States and various other countries. The list is shown in Sub-Appendix A.

The Martino Technology Concept:

The Martino Technology can be readily envisaged through the use of two diagrams. The first is shown on the next page is taken from the patent application of the first filing on May 19th, 1995. It is Figure 1 of that disclosure and depicts the three-tier system. From the text, the explanation is: FIG. 1 is a schematic diagram of a system 10 for entering data transactions into databases in accordance with the invention. As illustrated, system 10 comprises a first tier for capturing a data transaction having a one-to-many relationship to file structures, a second tier for exploding the data transaction into component parts having a one-to-one relationship to file structures, and a third tier for providing additional explosion of the data transactions for specific applications.

Figure 1 is from the Patent filing. Figure 2 is an actual system proposed for the military. This shows the relationship of the same multi-tier systems of Figure 1 with the imposition of terminology specific to the application. Interjected are the terms UNI-LINK® for the entire system, CYBERAGE® – Application Generator Engine for the AGE thin client, and CYBERBOSS® – Broad Operations System Server for the BOSS control network. These terms are trademarked by CyberFone Technologies, Inc, CFTI, the assignee of all the patents and other intellectual property of Dr. Martino.

Cyber Technology Group, Inc. a partner of CyberFone Technologies, Inc. has established a Beachhead for application development and professional services.

The example shown in Figure 2 is directed to a specific application area which depicts the use of UNI-LINK® for integrated military operations. What is significant is the interoperability of the thin-client AGE in a variety of end user devices – weapons, communications gear, monitoring, etc. In all cases, the AGE-BOSS relationship can provide security from penetration within a framework of continuous control of operations – all in real-time.

U.S. Patent Sep. 8, 1998 Sheet 1 of 12 **5,805,676**

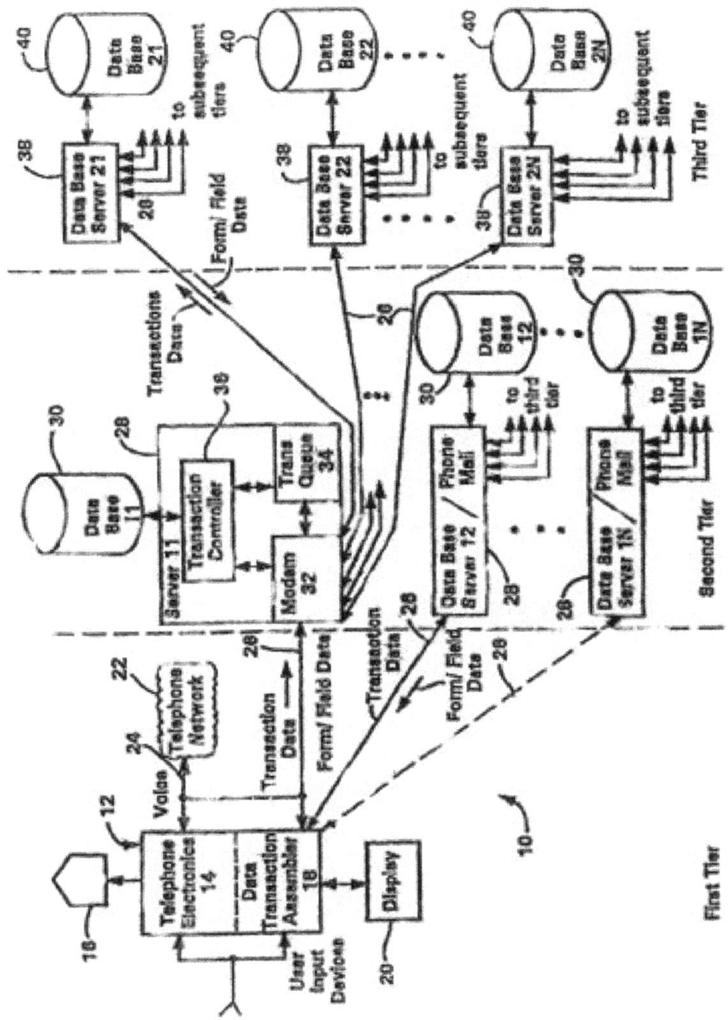

Figure 1: Martino US Patent 5,805,676

While this depiction includes potential hazards faced by the combatants, it must be stated that the UNI-LINK® system operates on-demand as well as in a reactive control mode. In whatever capacity or application that the system addresses in real time, it should be understood that this interoperable logic-based system can just as easily be applied to other governmental and NGO agencies as well as commercial enterprises responsible for dealing with man-made and environmental disaster planning. The UNI-LINK® system can create broadcast or targeted alerts providing individual or group determined protocols for expected response. This constitutes delivering context sensitive information to the point of need. It can also, of course, be used for such needs as instant messaging, multi-media communication, balloting, and entertainment.

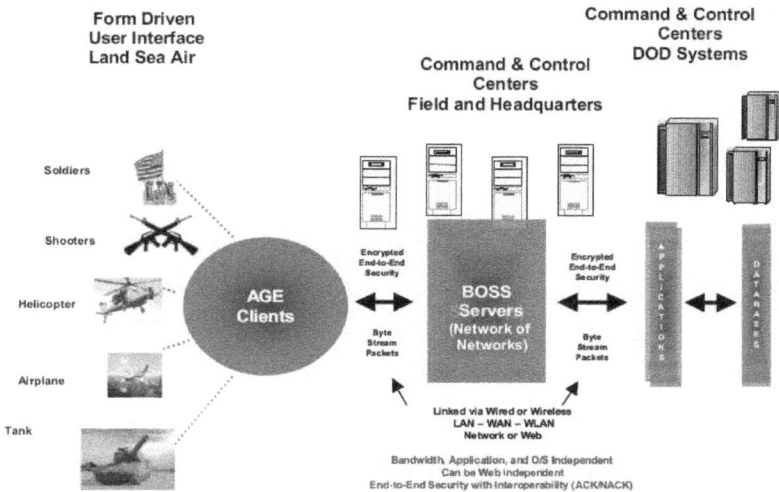

Figure 2: The **UNI-LINK®** System as applied to military operations

An example of the universal applicability of the Martino IP Technology

Advantages of the Martino Technology

As already stated, the Martino Technology is directed to convergent technology, unifying computer power, universal communication capability, and visualization technology. The Martino Technology also provides a set of tools for rapid wide-ranging application building and significant ease of maintenance. Using the Martino technology, systems and applications can be utilized in a vast array of devices, independent of variations in operating systems and hardware capability.

From a marketing and profit point of view, applications become:

1. In the forefront of modern convergent technology. Devices, services, and applications can be developed and marketed to meet the instant multi-media demands of the present and future – in a host of disciplines. Furthermore, these products can be marketed to function with currently available devices and services.

2. Relatively easy and quick to build. Employing the core UNI-LINK® technology that consists of the thin client Application Generator Engine (AGE) and the controlling middleware Broad Operation System Server (BOSS), creates an integrated, seamless, secure, real time front–end to back-end system. This patented core system provides for linkages to grids of networks for storage, extraction and updating without requiring changes in disparate database structures. For the user, the unity and simplicity of the AGE/BOSS UNI-LINK® system enables rapid application development that does not require programming skills. The companion Application Creator tool enables this rapid, inexpensive application development.

3. Operating System and hardware agnostic. The Transaction Assembly or Application Server is the thin client module (operating system) that

directs the microprocessor to execute in accordance with the requirements of the parameters of a form/menu that define a specific transaction type. Any other operating system present in the device is not necessary to execute transactions based on UNI-LINK® technology. Since both push and pull functionality are provided in this patented system, device hardware features can be as varied as the transaction type(s) to be employed may require, but the operating system principles are the same regardless of device type.

4. Materially simpler to maintain – O/S independent, changed at the logic level only, immediately tested and implemented. Using a tool-set created by CFTI, forms layouts can be developed pictorially showing the 'form'. This can be created by end-users and need not involve such end-users instructing experts of what is desired and needed. Hence origination and changes are made at the logic level by knowledgeable users and not be remote specialists. However, it must be noted that experts may be needed for special situations. That has always been true, and undoubtedly will continue to be true.

However, the need for such experts will certainly be less than current levels of use and experiences.

5. Capable of functioning on devices with minimal hardware resources. Nanotechnology inspired devices, wrist or other wearable devices or environmental control devices are a few examples of this utility.

6. Capable of working wired or wireless, on the Internet or not. The UNI-LINK® technology is network agnostic and therefore provides for singular or convergent two-way communication across communication networks

7. Language independent. Since forms and prompts are created by the end-user, with or without specialized assistance in special situations, the language readily varies for any end-user vernacular. Such changes can then be electronically updated. In the same fashion, the impact of changes can be minimal except for any impacts on database structures, which can be significant. Even in such cases, however, special tools have been created in the past to handle such changes.

8. Secure. As included in the patents, encryption techniques can be incorporated within UNI-LINK®. UNI-LINK® relies on the use of both industry standard and proprietary Martino encryption techniques. In the past such techniques were developed at XRT to securely transfer billions of dollars daily to corporations,

governments, and institutions worldwide. In over 20 years employing these techniques no loss of transferred funds was ever identified.

9.   Multi-media. The forms-based system provides for the transmission of any medium or media, either static or streaming, singularly or in combination, that is defined as digitalized data.

10. Through AGE, UNI-LINK® can interface with any existing system on any devices under any O/S. This can include cameras; wristwatches; and even nanotechnology-based devices that can reside within containers and bodies, including the human body.

The Universality of UNI-LINK® Application Potential

The Martino IP is protected by patents granted in various countries around the world. This technology serves to capture, receive and send, store, and update databases or communicate with other user communication devices

The system can be applied to an almost limitless number of transaction types including the several examples below:

1.   Audio and video data streams (pictures, graphics, medical images, movies, songs,

2.   VoIP (Voice over IP)

3. Health monitoring and control systems (health monitoring sensors (heart rate, temperature, blood flow velocity)

4. Financial transactions (home banking, securities trades) account lookup, transaction execution, account updates, reports, audit log

5. Military Command and Control systems (remote land, sea air)

6. Retail manufacturers rebate processing at the point of sale

7. Remote Learning (tutorials, simulations, testing, collaborative learning)

8. Secure local and remote electronic voting systems

9. Man-made and natural disaster and pandemic planning, training, responding, recovery

10. Pandemic response systems

11. Environmental data captured by monitoring sensors (liquid, gas, heat, cold)

12. Land and sea Navigation systems

13. Pharmaceutical Clinical trials monitoring

14. Pervasive computing

15. Enterprise Continuity Planning

16. Ultra-low-cost laptop functionality

<u>Conclusion</u>:

With such universal applicability and ease of implementation, the potential impact can be to create new industries. Using the Martino Technology can generate significant revenues, all protected by patented claims. This technology is in the forefront of the modern wave of convergent technology with a priority date of May 19th, 1995.

Sub-Appendix A

CyberFone Patents Assigned to
CyberFone Technologies, Inc.

United States

US 1   5,805,676
Filed 5/19/1995

US 2   5,987,103
Filed 8/11/1997

US 3   6,044,382
Filed 6/20/1997

US 4   6,574,314
Filed 9/7/1999

US 5   6,973,477
Filed 6/7/2000

US 6   App 10/947,853
Filed 9/23/2004

US 7   7,334,024
Filed 2/10/2005

These patents (granted and pending) have a Priority
Date of 5/19/1995

US Divisional Applications

Attorney Docket No. 19047-002003
Filed 4/12/2007

No. 19047-002004
Filed 9/4/2007

No. 19047-002005
Filed 9/4/2007

No. 19047-002006
Filed 9/4/2007

No. 19047-002007
Filed 9/6/2007

No. 19047-002008
Filed 9/6/2007

No. 19047-002009
Filed 9/6/2007

International Portfolio

EPC

EPC 96915846.8   Filed 5/16/1996
Granted

EPC 98931240.0   Filed 6/22/1998
Allowed

Pub. No. 1052567A Filed 9/19/2003
Under Exam

Canada

2,221,853   Filed 5/16/1996
                    Granted

2,295,139   Filed 6/22/1998
                    Granted

2,411,457   Filed 6/7/2001
                    Granted

Mexico

978955   Filed 5/16/1996
                    Granted

9911824   Filed 6/22/1998
                    Granted

Taiwan

87109969   Filed 6/22/1998
                    Granted

Singapore

9906038-6   Filed 6/22/1998
                    Granted

Israel

133496   Filed 6/22/1998
                    Granted

155300   Filed 12/5/2005
                    Granted

Korea

7016744 Filed 10/02
Awaiting Exam

Hong Kong

EPC 96915846.8 Filed 5/16/1996
Granted

07103006.9 Filed 3/20/2007
Pending

Sub-Appendix B

Sample Application Types
Wired / Wireless

General functionality

1. VoIP, Set top Box Audio/Video Control, Real time secure payment systems, Simulations

2. Audible Menus converted to Visible Menu

3. Emergency Management/Homeland Security

4. Disaster Planning, Response, Recovery

5. Maritime Port Security

6. Integrated Airborne/Land-based wireless communication

Real time, Secure Transaction Processing

1. Banking

2. Remote patient monitoring, diagnosis, and prescription

3. Home security system and general appliance devices control

4. Lottery Wagering

5. Enhanced Remote Learning/Simulations

6. Voting Information Management

7. Video on Demand

8. Retail Instant Rebate and Customer Specific Marketing at the Point of Sale

Military

1. Navy on Ship Communications

2. Army Base and Field Operations / Advanced Weapons Command and Control

Sub-Appendix C

Patent Protection Verification of Selected Application Areas

Application	US Patent No.	Claim(s)	Disclosure	Comment
VoIP Messaging	5,805,676	15,16,19, 37,38	Figure 6, Figure 2 Field "N"	
			Abstract,	
			C1 L 11-14	
			C2 35-42	
			C3 51-54	
			C6 L 53-67	
			C7 L 1-6, L 35-43	
			C9 L 62-67	
			C12 L 42-54	
			C13 L 11-15	
			C26 L 60-68	
			C27 L27 1-14	
	5,987,103	All	See 11/07 Analysis	
	7,334,024	All		

Mobile Browser and Menu Download Systems on cell phones Movies, music, books, etc from remote server	5,805,676	All	C13 L No.2 Software Section	See claim 35 for portable phone
Touch Screen input	5,805,676	29	C3 L 42-46 C7 L 44-49	
Text Messaging – text	5,805,676	15,16,31, 32	General	
Video Messaging	5,805,676 5,987,103	15,16,31, 32 37,38	General	
Home -Alarm systems, monitoring, control	5,805,676 Div. Appl 2009	 1,2,3,4,5, 6,7,8	C23 L 1-56	
Banking, securities trades, account lookup, transaction execution, bill payment, account updates,	5,805,676 5,987,103	1,2,3,4,5, 6,7,8,9, 13	C23 L23-31C23 L1-56 C19 L 40-44 Figure 12	

Reports, audit log	6,044,382			
	6,574,314, B1	13		
Internet audit log.	Div.app.20 09			
Set Top Box/remote configured as transaction entry device	5,805,676 6,044,382	31,32,34, 35	Figure 5 C32 L 41-55	
Airline Reservations	5,805,676		C23 L 37-51	
	Div. 2009	9,10		

Medical Local and Remote Monitoring and Diagnostic Instruments	5,805,676 6,044,382	31,32,33	C22 L 17-52 Fig 10,13,14 C5 L28-37	
	Appl. 19047-016P01 19047-015P01	All All	All All	
Retail POS Rebates	6,044,382		C5 L 28-30	
Secure Wireless/Wired message transaction system	6,973,477 B1	31,32	C20 L 52-67	
Remote Networks of servers sharing information (Grid Computing)	5,808,676 6,973,477 B1	All	Figure 1 Figure 11-A	

Cell Phone access to server farms	Div. App 2004	All		
Real-time entry of opinion and transaction data on portable devices	5,805,676		Figure 1	
Linkage to an unlimited number of databases External to the primary client/server system	5,805,676 6.973.477 B1	13 All	Figure 1 Figure 11-A	
Cell Phone Voice Recognition	5,987,103 Div app.2006	42 All		
Interoperable common applications executed from cell phones,	5,805,676	All	Figure 1	

PDAs, Tablet PCs, Laptops, etc				
Multi-tier, multi-dimensional, multi-media systems	5,805,676		Figure 1	
Electronic voting systems (wired or wireless)	5,805,676		Figure 1	
Visual representation of Navigation options of phone mail menus	5,805,676		C23 L 57-67 C24 L 1-26	
	5,987,103	33,34		
	Div. app. 2009	11,12		

Appendix VII

United States Patent Number 5,805,676
Granted on September 8, 1998

US005805676A

United States Patent [19]

Martino

[11] **Patent Number:** **5,805,676**

[45] **Date of Patent:** **Sep. 8, 1998**

[54] **TELEPHONE/TRANSACTION ENTRY DEVICE AND SYSTEM FOR ENTERING TRANSACTION DATA INTO DATABASES**

[75] Inventor: **Rocco L. Martino**, Villanova, Pa.

[73] Assignee: **PCPI Phone, Inc.**, Villanova, Pa.

[21] Appl. No.: **446,546**

[22] Filed: **May 19, 1995**

[51] Int. Cl.⁶ **H04M 11/00**
[52] U.S. Cl. 379/93.17; 379/93.25; 379/93.01; 395/203
[58] Field of Search .... 379/89, 90, 94, 379/96, 98, 202, 88, 110, 93, 87; 364/400, 401 M; 381/110; 348/13, 14; 395/201–204

[56] **References Cited**

U.S. PATENT DOCUMENTS

4,591,662	5/1986	Legros et al.	179/2 DP
4,598,171	7/1986	Hanscom et al.	381/110
4,689,478	8/1987	Hale et al.	379/96
4,776,016	10/1988	Hansen	381/110
4,835,372	5/1989	Gombrich et al.	379/93
4,851,999	7/1989	Moriyama	364/400
4,858,121	8/1989	Barber et al.	395/202
4,860,342	8/1989	Danner	379/96
4,972,462	11/1990	Shibata	379/89
4,984,155	1/1991	Geier et al.	364/401
4,991,199	2/1991	Parekh et al.	379/97
5,008,927	4/1991	Weiss et al.	379/98
5,189,632	2/1993	Paajanen et al.	364/705.05
5,195,086	3/1993	Baumgartner et al.	379/202
5,195,130	3/1993	Weiss et al.	379/98
5,301,105	4/1994	Cummings, Jr.	364/401 M
5,333,266	7/1994	Boaz et al.	379/89
5,351,076	9/1994	Hata et al.	348/14
5,365,577	11/1994	Davis et al.	379/96
5,416,831	5/1995	Chewning, III et al.	379/96
5,572,421	11/1996	Altman et al.	395/203

OTHER PUBLICATIONS

"a New Generation Of Information Terminals"Ceremtek Electronics Brochure, Apr. 1987.

Primary Examiner—Curtis Kuntz
Assistant Examiner—Stephen W. Palan
Attorney, Agent, or Firm—Woodcock Washburn Kurtz Mackiewicz & Norris LLP

[57] **ABSTRACT**

A data transaction processing system in which transaction data is entered by the user in response to prompts in a template which is tailored to each user application. The template and entered data are accumulated into data transactions which are immediately transmitted upon completion to an external database server for processing and storage. The data transactions are not locally stored for processing, and no conventional operating system is necessary. No local processing needs to be provided, and the only local storage is a flash PROM which stored the control firmware, a flash memory which stores the data streams making up the forms and menus, and a small RAM which operates as an input/output transaction buffer for storing the data streams of the template and the user replies to the prompts during assembly of a data transaction. The data transaction is received via standard protocols at a database server which, depending upon the application, stores the entire data transaction, explodes the data transaction to produce ancillary records which are then stored, and/or forwards the data transaction or some or all of the ancillary records to other database servers for updating other databases associated with those database servers. Also, in response to requests from the transaction entry device, the database server may return data streams for use in completing the fields in the data transaction or in presenting a menu on the display which was read in from the database server or a remote phone mail system. The transaction entry device is integrated with a telephone and is accessed via a touch screen, an optional keyboard, a magnetic card reader, voice entry, a modem, and the like.

44 Claims, 12 Drawing Sheets

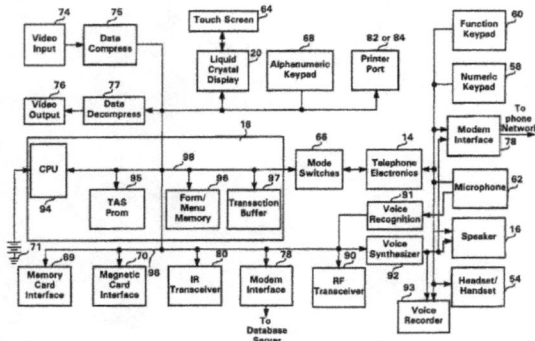

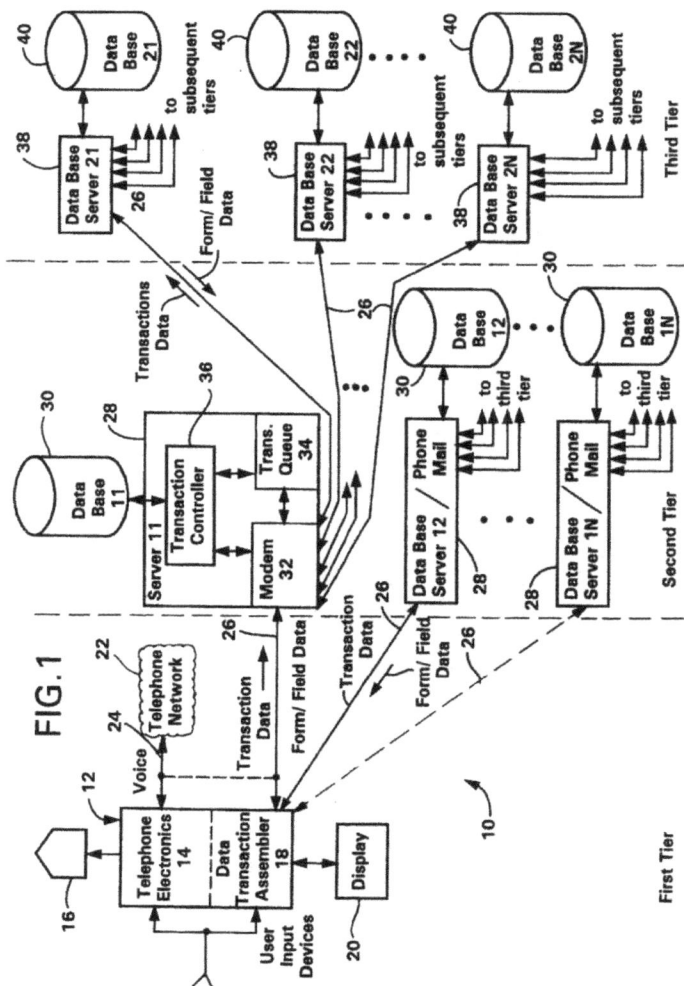

FIG.1

FIG.2

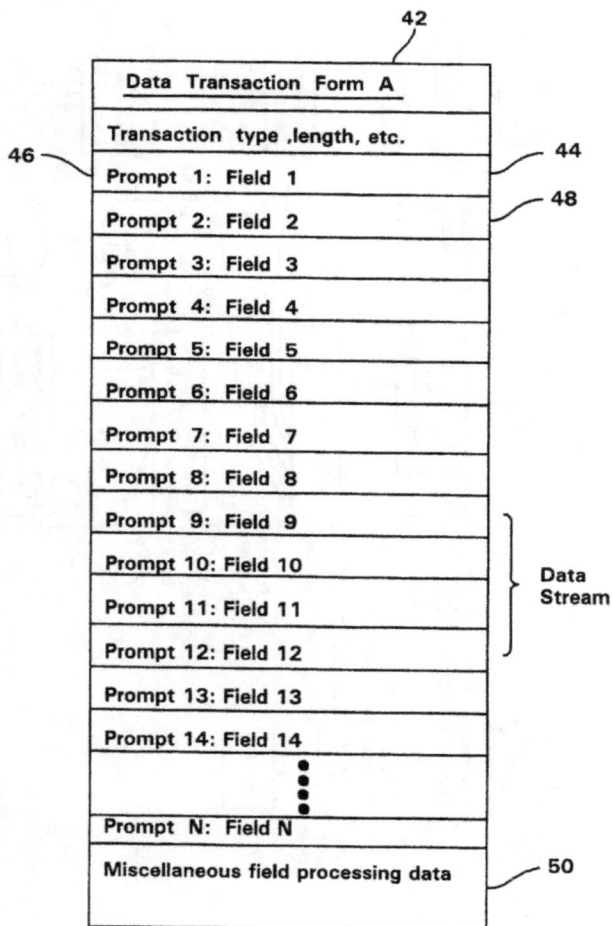

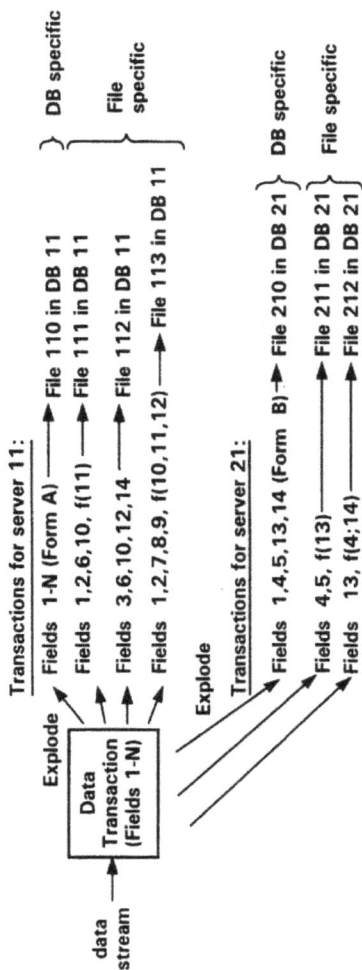

FIG.3

FIG.4

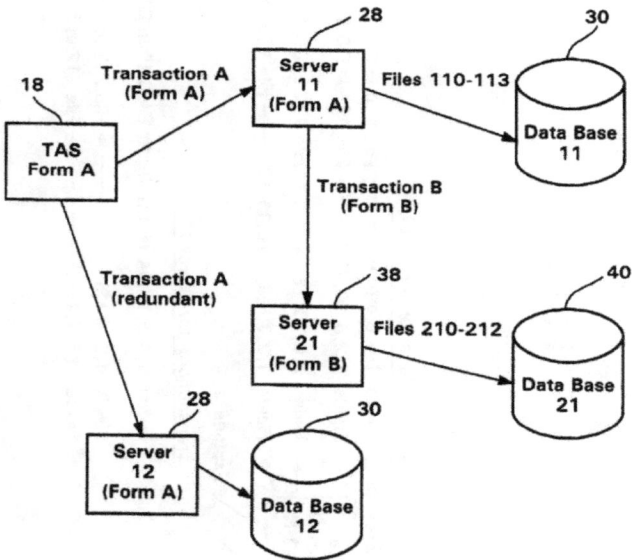

244

U.S. Patent   Sep. 8, 1998   Sheet 5 of 12   **5,805,676**

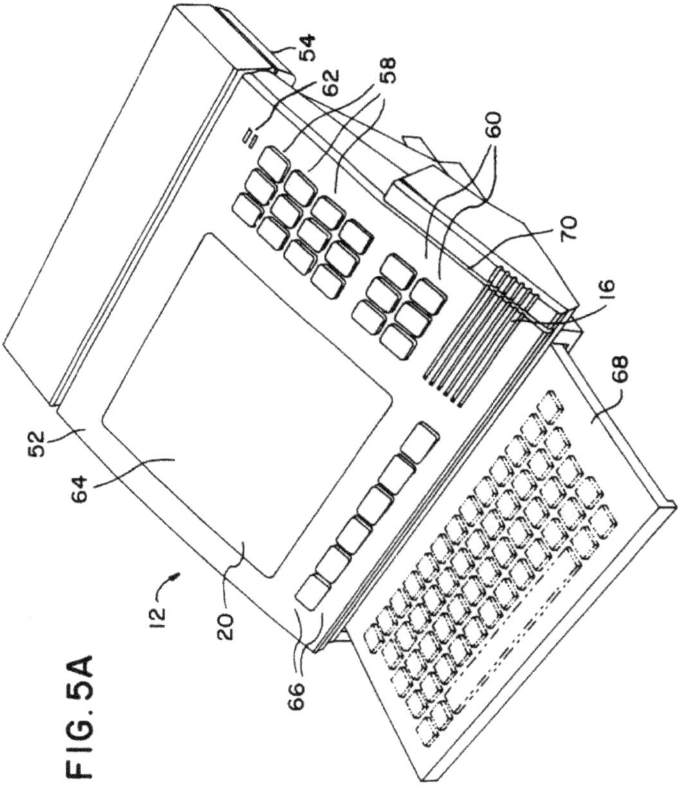

FIG.5A

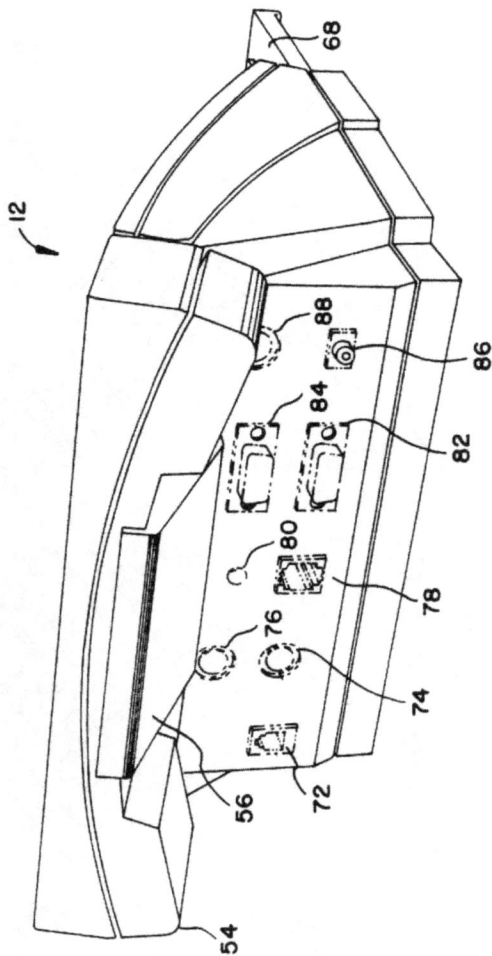

FIG. 5B

246

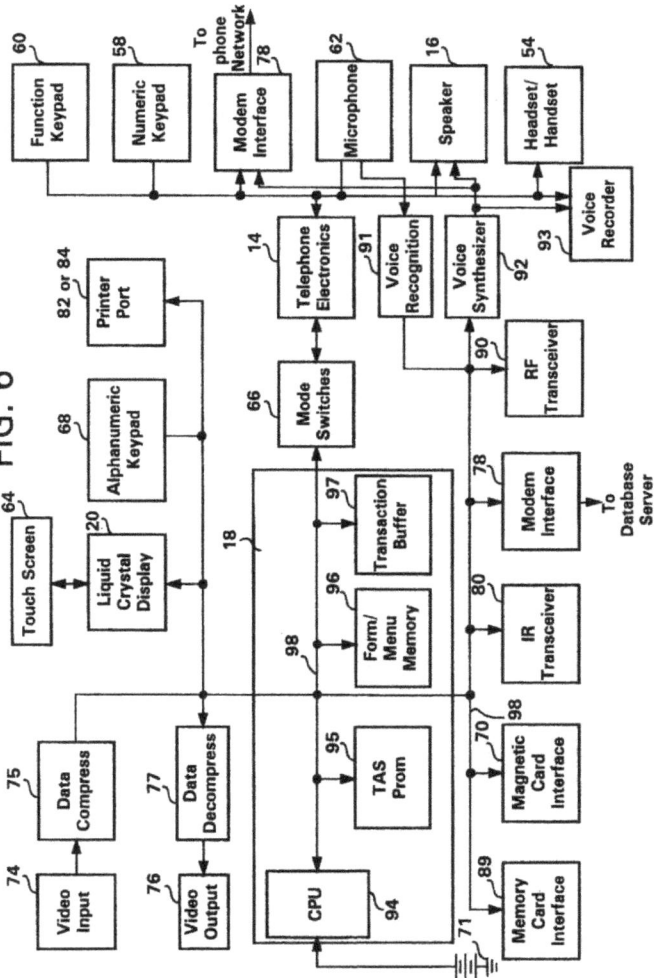

FIG. 6

FIG.7

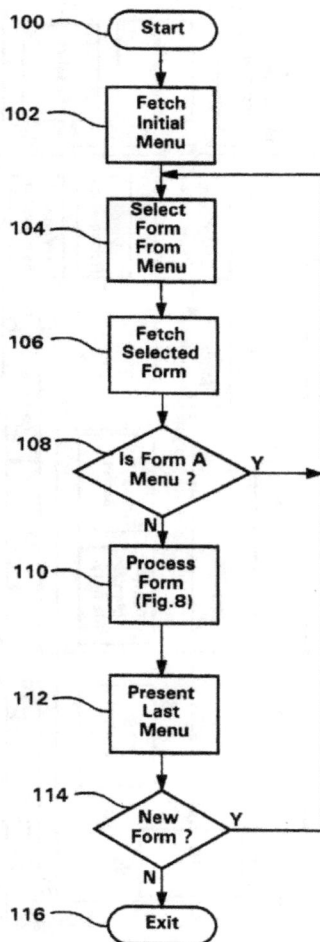

TAS

100 — Start

102 — Fetch Initial Menu

104 — Select Form From Menu

106 — Fetch Selected Form

108 — Is Form A Menu ?   Y

N

110 — Process Form (Fig.8)

112 — Present Last Menu

114 — New Form ?   Y

N

116 — Exit

U.S. Patent   Sep. 8, 1998   Sheet 9 of 12   **5,805,676**

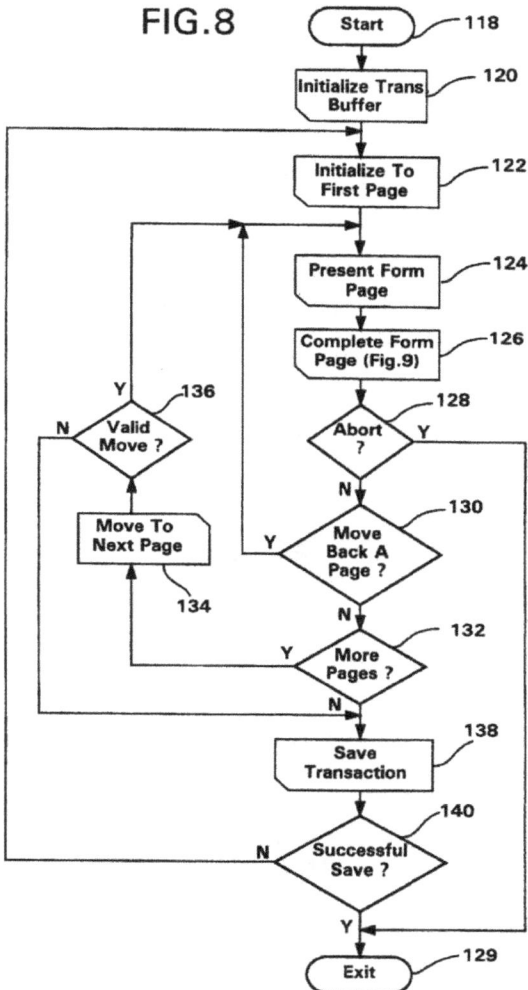

FIG.8

Start — 118

Initialize Trans Buffer — 120

Initialize To First Page — 122

Present Form Page — 124

Complete Form Page (Fig.9) — 126

Abort ? — 128

Valid Move ? — 136

Move To Next Page — 134

Move Back A Page ? — 130

More Pages ? — 132

Save Transaction — 138

Successful Save ? — 140

Exit — 129

249

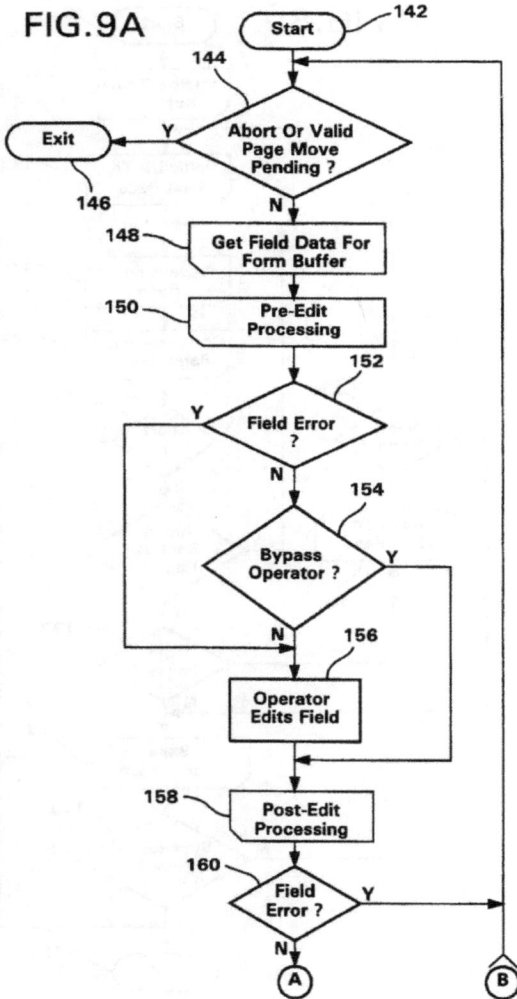

FIG.9A

FIG.9B

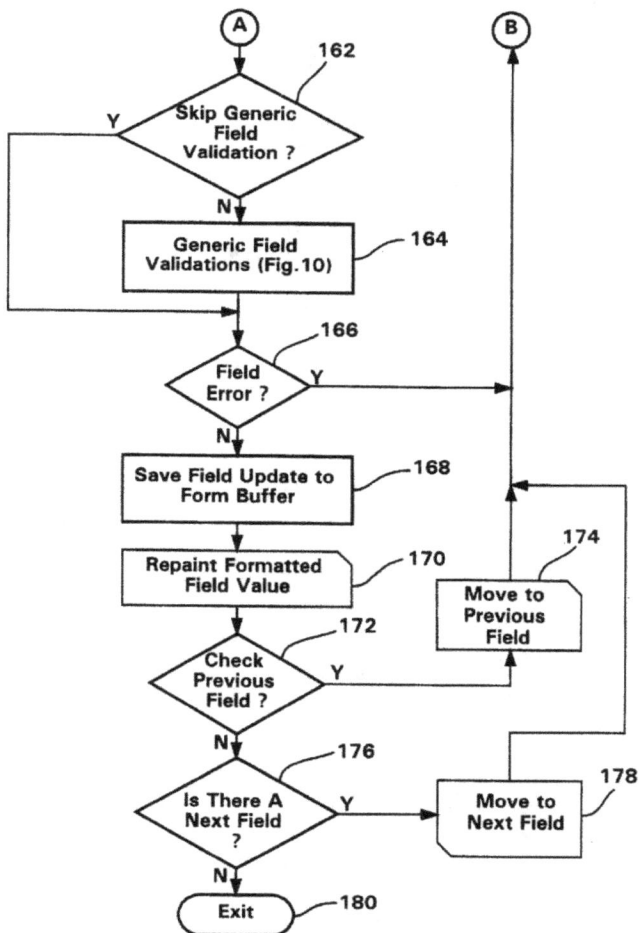

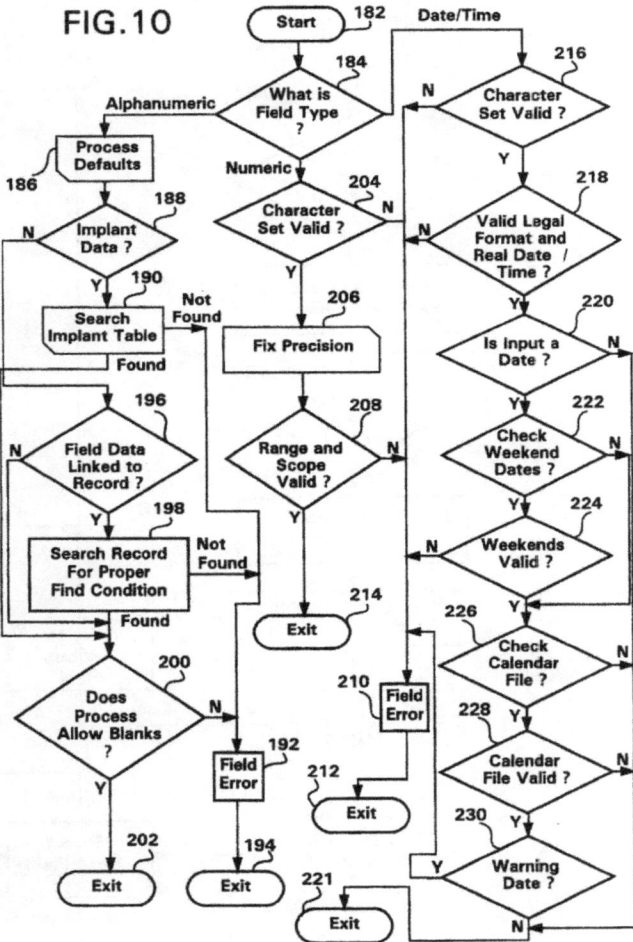

FIG.10

5,805,676

| 1 | 2 |

TELEPHONE/TRANSACTION ENTRY DEVICE AND SYSTEM FOR ENTERING TRANSACTION DATA INTO DATABASES

BACKGROUND OF THE INVENTION

1. Field of the Invention

The present invention relates to a system for automatically capturing data at the point of transaction and storing that data in the appropriate database(s), and more particularly, to a data transaction processing system including a transaction entry device which can selectively operate in a telephone mode and a transaction entry mode. In the telephone mode, the transaction entry device operates as a conventional telephone. However, in the transaction entry mode, menus are used to navigate the user to forms which facilitate the entry of data. The entered data and forms together form data transactions which are transmitted to one or more databases for processing and storage. The database(s) also "explodes" the data transactions into their component parts and transmits those component parts to still other databases for processing and storage so that the data in the data transactions automatically updates all current database items affected by such data.

2. Description of the Prior Art

The telephone has become an increasingly versatile instrument. The functionality of telephones has been expanded by incorporating the functions of answering machines, facsimile machines, and the like. Point-of-entry systems have also been developed which incorporate computer processing capabilities into conventional telephones. For example, a computer/telephone apparatus is described in U.S. Pat. Nos. 5,195,130, 5,008,927, and 4,991,199 which configures a telephone as a programmable microcomputer which is operated through the standard telephone 12-key keypad. A programmable gate array is reconfigured to accommodate various types of software which require different hardware configurations but without actually reconfiguring the hardware. The reconfiguration data is received from a network host computer and is used by the programmable microcomputer to emulate the hardware of any of a plurality of service bureaus which communicate with the network host computer. In this manner, the telephone/computer is configured to communicate data to/from any of a number of different service bureaus via conventional telephone lines.

However, telephone/computer systems of the type described in the afore-mentioned patents are typically quite complicated and expensive and are limited by the types of operating software which can be downloaded from the network host computer. Also, such telephone/computer systems are relatively slow since the microcomputer must be reconfigured before it will permit communication with the requested service bureau. Because of these characteristic features, such telephone/computer systems are typically used in public locations and are not efficient for creating point-of-entry transactions in typical commercial or private settings. A point-of-entry transaction system is desired which does not have such limitations and which is operating system independent.

Elimination of the requirement of a conventional operating system and the associated application programs for the microcomputer of a data entry device would greatly decrease the cost of such a device. However, to date, this has not been possible because the operating system is needed to run the application programs which control the data communications and together handle discrete parts of the sys-

tem. Unfortunately, such application programs require substantial amounts of local memory and substantial processing power for performing the desired functions. Also, the operating systems themselves tend to be quite costly to purchase and maintain.

Accordingly, a data entry system is desired which does not have the inherent limitations of conventional point-of-entry systems such as the requirement of a standard operating system for communication with a remote service bureau or file server. A data entry device and associated system is desired which performs a minimal amount of processing at the data entry device so that the data entry device may be as simple and inexpensive as possible, thereby bringing the cost of such a device into a range suitable for most commercial and private uses. It is also preferable that such a data entry device provide a wide range of functionality without requiring a local operating system program and a plurality of applications programs for implementing each function. The present invention has been designed to meet these needs.

SUMMARY OF THE INVENTION

The system which meets the above-mentioned needs in the art includes a transaction entry device that permits the user to organize and control all aspects of his or her personal transactions as well as any transactions that may occur in an office setting. In its simplest terms, the transaction entry device formats input data into a data transaction having content which is dependent upon the type of application to which the associated data pertains. These data transactions are then transferred to a local or remote database server which "explodes" each data transaction into its component parts for updating all databases containing data to which the data in the component parts pertain. In this "transaction entry mode" the transaction entry device of the invention functions as a multi-purpose workstation. However, since the data transactions are created without the use of an operating system or application programs, the transaction entry device is quite simple and inexpensive and may be readily integrated with the customer's desktop telephone or portable telephone.

The present invention combines computer technology and telephone technology to allow transaction data to be captured at the point of initiation of the transaction. The transaction entry device is integrated into a conventional telephone which acts as either a normal telephone in a telephone mode or as a transaction entry device in a transaction entry mode. When in telephone mode, the telephone operates in a conventional manner. However, when in transaction entry mode, the transaction entry device is driven by a microprocessor which is, in turn, driven by an operating system independent transaction assembly (or application) server (TAS) comprised of data streams stored in a flash PROM. The TAS is absolutely self-contained in its relationship to the hardware of the transaction entry device and in general performs the two basic functions of (1) generating a template or form from a data stream and (2) developing a data transaction as the user inputs data in response to prompts in the template or form. The template is a series of data streams read from a local flash memory or transmitted directly from an external source such as a database file server.

During operation, the data entered by the user in response to prompts in the template are accumulated into data transactions which are immediately transmitted to an external database server. Unlike typical prior art systems, the data

5,805,676

3

transactions are not locally stored for processing by the local microprocessor once the data transaction has been completed. On the contrary, the only required storage in the transaction entry device is a flash PROM for storing the TAS firmware, a flash memory for storing the data streams used by the TAS firmware to complete a form and the modem numbers for the remote database servers, and a small RAM which operates as an input/output transaction buffer for storing the data streams of the template and the user replies to the prompts in the template during assembly of a data transaction. The transaction buffer(s) only needs to be as large as the largest data transaction since it only stores the form until the entire data transaction is completed. In this sense, the transaction entry device serves as an assembly point for specific transactions until they are ready for transmission to an external database server for processing and storage.

The data transaction formed by the transaction entry device is transmitted via modem to a local or remote database server for processing and storage. The data transaction is received via standard protocols at the database server which, depending upon the application, stores the entire data transaction, explodes the data transaction to produce ancillary records which are then stored, and/or forwards the data transaction or some or all of the ancillary records to other database servers for updating other databases associated with those database servers. Also, in response to requests from the transaction entry device, any of the database servers may send data streams back to the transaction entry device for use in completing the fields in the data transaction or in displaying new forms or menus for selection.

Thus, the data transaction system of the invention comprises at least three tiers: a first tier for capturing the data transaction from the user, where the data transaction has a one-to-many relationship to file structures; a second tier for exploding the data transaction into its component parts on a system-specific basis so that each component part has a one-to-one correspondence with a file; and a third tier for providing additional explosions of the data transactions on an application-specific basis so that each application has its own set of data transactions.

A preferred embodiment of the transaction entry device of the invention resembles a conventional telephone except that it includes a touch screen and an optional keyboard for data entry in addition to the conventional numeric and function keypad inputs. A telephone handset or headset is optional and may be replaced by a microphone and speaker. The transaction entry device of the invention also includes RS-232 and other input/output ports for accommodating other options such as a wireless (RF) receiver, a magnetic card and/or smart card reader, a video camera and video display, infrared controllers, and the like. The telephone preferably has normal touch-tone functions as well as mobile and cellular options.

Preferably, the transaction entry device contains a microprocessor such as an Intel 80386SX or higher, one megabyte of flash memory for dynamically storing the data streams for the templates, one megabyte of flash PROM for storing the TAS firmware, and a 128 kB RAM which functions as a transaction buffer for storing the data streams of the templates and the user responses until completion of the data transaction. A graphics display screen is also provided for displaying the templates to the user for the entry of the data transactions. Preferably, the graphics display screen is on the order of 24 lines by 40 characters for a desktop unit and 12 lines by 40 characters for a cellular unit.

4

The transaction assembly (application) server (TAS) guides the user to the desired template via menu selections, where the menus and templates are stored in flash memory and are called up by the TAS firmware when selected by the user. Generally, the menus are treated as a special type of template or form. The templates stored in the flash memory may be updated at any time to handle particular applications, the user applications may be generalized so that no unique user application programs need to be written when a new application is added. However, if code is needed, or if a multimedia element is to be included in a data transaction, it can be appended to a data transaction as an additional parameter stream in the stream of data forming the data transaction. Also, since the nature of the data in the respective fields of the templates for particular applications is known in advance, the interface to a database server to permit storage of the data transactions and their component parts in the appropriate databases in the appropriate formats for each database becomes trivial.

In an alternative implementation of the invention, a process may be selected from the menu of the transaction entry device which creates a "visible" menu corresponding to a voice mail menu of a remote phone mail system. When such a process is selected, the telephone or modem interface makes a telephone connection to the remote phone mail system, and, once the connection is made, the data transaction assembler sends a data request for a visual representation of the phone mail menu of the remote phone mail system via the telephone connection to the remote phone mail system. A data stream containing the visual representation of the phone mail menu from the remote phone mail system is then returned via the telephone connection and stored in a memory of the transaction entry device for presentation to the display screen of the transaction entry device 12. When the desired phone mail menu option is selected from the "visible" voice mail menu, the data transaction assembler creates a data transaction indicating which menu item was selected and sends the data transaction to the remote phone mail system via the telephone connection. Based on the menu selection, the remote phone mail system then returns a data stream containing a visual representation of the next phone mail menu via the telephone connection for storage and display. This process is repeated until the calling party is required to leave a message or the called party is reached.

BRIEF DESCRIPTION OF THE DRAWINGS

The above-mentioned characteristic features of the invention will become more apparent to those skilled in the art in view of the following detailed description of the invention, of which:

5,805,676

| 5 | 6 |

FIG. 1 is a schematic diagram of a system for entering data transactions into databases in accordance with the invention.

FIG. 2 illustrates a generic template for use in creating a data transaction in accordance with the invention.

FIG. 3 illustrates an "exploded" data transaction in which the component parts of a data transaction are stored in database-specific and file-specific locations.

FIG. 4 illustrates the "exploded" transaction of FIG. 3 in the context of the system illustrated in FIG. 1.

FIGS. 5(a) and 5(b) together illustrate a preferred embodiment of a transaction entry device in accordance with the invention.

FIG. 6 is a schematic diagram of the electronics of the transaction entry device illustrated in FIGS. 5(a) and 5(b).

FIG. 7 is a flow diagram of a menu driven transaction assembly (application) server (TAS) in accordance with the invention.

FIG. 8 is a flow diagram illustrating a technique for processing a form used to form a data transaction in accordance with the invention.

FIGS. 9(a) and 9(b) together illustrate a flow diagram of a technique for completing and editing a data transaction in accordance with the invention.

FIG. 10 is a flow diagram illustrating how the TAS validates the fields of each data transaction.

DETAILED DESCRIPTION OF THE
PRESENTLY PREFERRED EMBODIMENT

A system and method which meets the above-mentioned objects and provides other beneficial features in accordance with the presently preferred exemplary embodiment of the invention will be described below with reference to FIGS. 1–10. Those skilled in the art will readily appreciate that the description given herein with respect to those figures is for explanatory purposes only and is not intended in any way to limit the scope of the invention. For example, those skilled in the art will appreciate that the telephone/transaction entry device and system for entering data transactions into remote databases in accordance with the invention may be used in numerous settings in numerous applications. Accordingly, all questions regarding the scope of the invention should be resolved by referring to the claims.

A. SYSTEM DESCRIPTION

1. Overview

The system of the invention provides for the automatic capture and computerization of data associated with data transactions as they occur. As used herein, a data transaction is the combination of a form or template or a series of forms or templates containing data entry prompts and the data entered in response to those prompts. Throughout the remainder of this specification, the words "form" and "template" will be used interchangeably.

The data transactions are generated by a transaction entry device through an interactive process between the user and the form. The data transaction is assembled in a transaction buffer in the data transaction entry device and then transmitted to an external database for storage. No local storage for data transactions is available. The data transaction is defined externally by the database in that all applications consist of a series of customized forms and prompts for soliciting entry of the data needed to update the databases containing data related to the particular application.

Generally, the data transaction will have a one-to-many relationship to the file structures of the database containing data for that application.

The data transactions are entered using the transaction entry device. Preferably, the transaction entry device is integrated with telephone electronics so that the resulting device may selectively operate as a conventional telephone or as a data transaction entry device. The resulting transaction entry device preferably includes a touch screen and/or keyboard which provides input to a transaction assembly (application) server (TAS) which, in turn, presents selection options via menus and forms for completion by the user. Menu and form selection and form completion is made by touch, by key selection from the keyboard, by moving a cursor to the appropriate selection point and depressing a key, or even by voice command. Whenever data entry (other than mere selection) is desired, it is accomplished via a menu-driven selection process and/or by direct entry of data using a keyboard, a keypad, a touch screen, and the like. In the menu-driven case, a set of options is presented to the display screen by the TAS firmware. If this set of options exceeds the capacity of the display screen, then the list is scrolled up or down through the use of scroll keys on the device, by voice command, or by touch at scroll command points. Once the selection is made, the data associated with that selection is automatically entered into the form from a local or remote database, or the data is input by the user. In the event of keyboard entry, the TAS firmware may present a keyboard at the bottom of the display screen for touch entry; alternately, an optional keyboard located at the base of the transaction entry device may be used.

When the data is entered independently of a selection process, such data also may be entered using a swipe card if the data resides on the swipe card or the data may be transferred into the data transaction via modem from an external source. The data read from the swipe card can be used to fill out a form or may be transmitted to an external database or computer. Data returned from the external database or computer via modem may also be used to fill out the fields in the form. As desired, the data in a data transaction may also be written to a swipe card or memory card and the like.

The TAS firmware of the invention stores the options as well as control programs (microcode) for the processor for use with the templates in creating the data transactions. The TAS firmware also includes a program allowing connection via modem to one or more external computers and databases. Preferably, two modes of operation are available: transaction entry mode (with or without modem connection) and telephone mode. A selection of either the transaction entry mode or the telephone mode is made through a switch selection on the transaction entry device.

When the transaction entry device is placed in the transaction entry mode, the TAS firmware immediately presents a selection menu for all of the options the system is programmed to handle. In the telephone mode, on the other hand, a dial tone is provided and the telephone keypad is enabled. In telephone mode, one or more lines may be connected so as to allow simultaneous use of the transaction entry device without interfering with the modem connection. However, if a single telephone line is used, the telephone capability is available at all times or intermittently via modem as specified by the particular application. In the intermittent mode, upon a "save" the transaction entry device will control a dial up and transfer of data to a remote database server. On the other hand, if the telephone is used with an automatic dialer mechanism utilizing a phone list,

5,805,676

7

the transaction entry device may automatically change from the telephone mode to the transaction entry mode. In this case, a display on the telephone may be used to present a name and telephone list from which a selection can be made in accordance with the menu selection techniques described below.

2. Data Transaction System (FIGS. 1–4)

FIG. 1 is a schematic diagram of a system 10 for entering data transactions into databases in accordance with the invention. As illustrated, system 10 comprises a first tier for capturing a data transaction having a one-to-many relationship to file structures, a second tier for exploding the data transaction into component parts having a one-to-one relationship to file structures, and a third tier for providing additional explosion of the data transactions for specific applications.

The first tier comprises a transaction entry device 12 which captures the data transaction from the user in response to any of a plurality of inputs from the user. Transaction entry device 12 includes conventional telephone electronics 14 and speaker 16 and a data transaction assembler 18 for creating a data transaction in accordance with the invention. A display screen 20 is preferably associated with data transaction assembler 18 so that the user may monitor creation of each data transaction. Telephone electronics 14 are connected to a telephone switching network 22 via a conventional voice connection 24 over the telephone lines, while data transaction assembler 18 is connected via telephone lines 26 to one or more database servers 28. As illustrated in FIG. 1, telephone lines 24 and 26 may be separate lines, thereby permitting simultaneous use of the telephone and data entry functions, or the telephone electronics 14 and data transaction assembler 18 may be connected to a single line as illustrated in phantom in FIG. 1. Of course, when the telephone electronics 14 and data transaction assembler 18 are connected to a single line, a mode switch will enable their mutually exclusive operation, or alternatively, any of a number of conventional transmission schemes may be used to permit simultaneous transmission of the voice from the telephone electronics 14 and the data from the data transaction assembler 18 over the same line.

During operation in the transaction entry mode, transaction entry device 12 is responsive to user input devices such as a touch screen, a telephone keypad, a keyboard, a microphone, a swipe card, a memory card, video input, and the like, to form data transactions using data transaction assembler 18. Alternatively, the transaction entry device 12 operates in a telephone mode as a conventional telephone and receives inputs from a microphone and/or a handset, a touch tone keypad, and the like. More details of the transaction entry device 12 and data transaction assembler 18 will be provided in the next section with respect to FIGS. 5–10.

The second tier comprises one or more database servers 28 and their associated databases 30. In general, each database server 28 receives data transactions from one or more transaction entry devices 12 and "explodes" the received data transactions into their component parts for storage in the appropriate files of the associated database 30. In other words, the one-to-many file structure of the data transactions from one or more transaction entry devices 12 is converted into many one-to-one data transactions for storage in individual files of database 30.

Each database server 28 includes a modem 32 for transmitting/receiving data from the telephone lines 26, particularly the data transactions from one or more transac-

8

tion entry devices 12. Preferably, the data transactions are transmitted over the telephone lines 26 as data packets having, for example, 128 bytes, where 120 bytes contain information and 8 bytes contain control data. A transaction queue 34 acts as an input buffer for the received data transactions and controls the rate of presentation of the data transactions to transaction controller 36. Transaction controller 36 processes the received data transactions to extract the physical file relationships of the component parts of the data transactions and stores the components parts and different combinations thereof in the appropriate files of associated database 30. Alternatively, transaction controller 36 may process a data request from data transaction assembler 18 requesting information from database 30 for completing certain fields of a data transaction being processed by the transaction entry device 12. Database 30 then provides the requested information to database server 28 which, via modem 32, provides a data stream back to data transaction assembler 18 for use in completing the data transactions or presenting additional menus and forms in accordance with the invention. Typically, a user ID and password are transmitted to the transaction controller 36 to permit a connection to be made by data transaction assembler 18. Thus, transaction controller 36 also checks and stores startup and logoff information in addition to storing data transactions and directing reconstituted data transactions to other database servers as described herein. In addition, database server 28 may include a conventional phone mail system with an associated database for storing voice mail messages. In this case, the data transaction may include voice data for storage in the remote voice mail system.

As shown in FIG. 1, several database servers 28 may be provided. Preferably, each transaction entry device 12 has an associated database server 28 for performing any desired processing of its data transactions, although it is preferred that the data transactions be copied to at least one other database server 28 as shown in FIG. 1. This redundancy minimizes the possibility of losing data in the event of a power outage and the like. Preferably, each database server 28 contains essentially the same hardware, although modem 32, transaction queue 34, and transaction controller 36 have not been shown for all database servers 28 for ease of illustration.

In the transaction entry mode, the data transaction assembler 18 of transaction entry device 12 creates a data transaction that is transmitted to an associated transaction controller 36 of an associated database server 28. By "associated" it is meant that the database server 28 functions to perform any processing requested or necessary in conjunction with the storage of a data transaction from a particular transaction entry device 12. Of course, a particular database server 28 may have several transaction entry devices 12 associated with it. So that no data will be lost, a particular database server 28 may also serve as a backup for another database server 28 in the event of the failure of any database server 28.

As will be explained in more detail below with respect to FIGS. 2–4, database server 28 "explodes" data transactions received from data transaction assembler 18 and provides the component parts of the "exploded" file dependent data transactions via modem 32 to other database servers 28 as necessary to update other databases. Alternatively, as shown by dotted line in FIG. 1, the "explosion" of the data transactions may be performed by the data transaction assembler 18 at the transaction entry device 12 and the component parts transmitted to all appropriate databases 28 for updating the data therein. For this purpose, the data

STAIRWAY TO THE SMARTPHONE

5,805,676

9**10**

transaction assembler **18** will also need to know the modem numbers for all database servers **28** to be updated by the exploded data transactions. However, those skilled in the art will appreciate that this latter alternative will require access to numerous phone lines by the transaction entry device and that such phone lines are not always available to the user.

Finally, the third tier of the system **10** includes additional database servers **38** and databases **40** which support file dependent data transactions for specific applications. This additional tier of database servers **38** and databases **40** permits the data in the data transactions to be routed to application specific databases for storage of application specific data and access by those transaction entry devices **12** requesting data related to that specific application.

The creation and storage of a data transaction in accordance with the invention will now be described with respect to FIGS. 2–4.

Data transactions are created by data transaction assembler **18** as a data stream of a known format. A generic data transaction is illustrated in FIG. **2**. As defined herein, a data transaction is created using a form containing one or more of the following: instructions, prompts, menu selection options, and a template with fields for data entry. Generally, the menu form consists of prompts for selecting a form, another menu, or a process, and a single slot for entering a selection, while the data entry form consists of prompts and instructions together with fields for entering data as shown in FIG. **2**. The data entry form can have either single or multiple fields for entering data.

In transaction entry mode, the user navigates through menus of data transaction assembler **18** until a form related to a particular type of data entry operation is selected. Once selected, data transaction form **42** is presented to the user on display device **20**. The data transaction form **42** is a collection of data defining the visual presentation on the display device **20** and a list of the fields through which linkages to external database files are defined.

As shown in FIG. **2**, data transaction form **42** includes a format field **44** which identifies the type of data transaction this form pertains to, the length of the form, the number of pages in the form, the number of bytes in each field, storage keys, and the like. The body of the data transaction form **42** comprises a predetermined series of prompts **46** which are provided to the display screen **20** as a data stream. The prompts preferably include descriptive data which may be alphanumeric, an icon, or a list that scrolls, if necessary. Fields **48** are blank spaces of predetermined size provided for accepting user input in response to each prompt. Generally, the size of each field **48** is also stored in the stream of data defining the data transaction form **42**. Since the prompts are tailored to elicit the necessary data for the application for which the data transaction form **42** was created, the fields **48** will include the user data necessary for processing a data transaction for that particular type of application. The user responses become part of the data stream which forms the data transaction. Typically, the data transaction form **42** also includes a miscellaneous processing field **50** which permits processing data unique to that form to be appended to the data transaction for transmission. Such processing data may include, for example, equations which define the relationships of the data in certain fields of the data transaction or audio or video data attached to a multimedia data transaction. In addition, non-display data associated with the time of data entry, the date of data entry, the user ID, and the like may be stored in miscellaneous processing field **50**.

FIGS. 3 and 4 illustrate the "explosion" of the stream of data forming the data transaction created using the data transaction form **42** of FIG. 2. As shown in FIG. 3, each data transaction contains data which is specific to a particular database and/or specific to particular files in one or more databases. The data in the data transaction is "exploded" accordingly. For example, the complete data transaction from FIG. 2 (Form A) is stored in a particular file (file **110**) of the database associated with the transaction entry device **12** which created the data transaction (database **11** in FIG. 1). Storage of the entire data transaction is desired so that records may be maintained in the event of system error, power failure, and the like. The transaction controller **36** then extracts data from those fields of the data transaction which it knows to be related in forms of that particular type. For example, the data in fields **1, 2, 6, 10**, and a function of the data in field **11** may relate to a particular application stored in file **111** of database **11**. Similarly, the data in fields **3, 6, 10, 12**, and **14** may be related to an application stored in file **112** of database **11**, while the data in fields **1, 2, 7, 8, 9**, and a function of the data in fields **10, 11**, and **12** may be related to an application stored in file **113** of database **11**. These fields are extracted from the received data transaction by transaction controller **36**, reconstituted into a file entry of the appropriate format (as necessary), and stored in the associated database **30**.

All of the data in the received data transaction, or a subset thereof, may also be retransmitted to one or more additional application specific databases, such as database **21** of the databases **40** in tier 3. As illustrated in FIG. 3, the database specific data of fields **1, 4, 5, 13**, and **14**, forming the subset (Form B) of the original transaction (Form A), is stored in file **210** of database **21** so that a complete record may be maintained. Subsets of the data in Form B are then stored in specific files of database **21** as indicated. In this manner, the data of the original data transaction (Form A) is automatically sent to all databases which contain files which must be updated by any or all of the data in Form A.

FIG. 4 illustrates the explosion of the data transaction in FIG. 3 for the system **10** illustrated in FIG. 1. As shown, the data in the data transaction (Form A) is extracted to update files **110–13** of database **11** as well as files **210–212** of database **21**. A redundant copy of Form A is also maintained in database **12**.

As will be explained more fully below, the system of FIGS. 1–4 is significant in that the data in a data transaction may update one or more databases serviced by file servers operating under control of numerous types of operating systems without the requirement of a terminal or operating system emulation by the transaction entry device **12**. On the contrary, the transaction entry device **12** of the invention permits data capture and storage with a minimum amount of processing at the transaction entry point (tier 1), which, of course, minimizes system cost.

B. Transaction Entry Device 12 (FIGS. 5–10)

As noted above, the transaction entry device **12** is particularly characterized by the data transaction assembler **18**, which controls the various operations of the transaction entry device **12** in its transaction entry mode. Preferably, data transaction assembler **18** uses simple menu structures and predetermined forms stored as data steams in a flash memory for facilitating data entry. The menus are treated as a special type of form and are used to call other menus, forms, or processes. The forms, on the other hand, are used to create data transactions which are sent to one or more file servers operating under different operating systems, where

5,805,676

| 11 | 12 |

the data transaction is "exploded" into its component parts for storage in a unique file structure for updating all records affected by the data in that data transaction. In turn, the "exploded" data transactions may be transmitted to another application specific database (tier 3) for storage. Processes, on the other hand, are selected to perform limited processing of the values in the fields of the forms. Such processing may be performed locally but is preferably performed by the associated database server 28.

1. Hardware

A preferred embodiment of a transaction entry device 12 incorporated into a conventional telephone is illustrated in FIGS. 5 and 6. As shown in FIG. 5a, a preferred desktop embodiment of a transaction entry device 12 includes a housing 52 on the order of 8 inches wide by 12 inches long for housing telephone electronics 14 and the hardware of data transaction assembler 18. Transaction entry device 12 includes an optional handset (or headset) 54, cradle 56 (FIG. 5b), numeric keypad 58, telephone function/line keys 60, microphone 62, and speaker 16, which facilitate operation of the transaction entry in the telephone mode. As known to those skilled in the art, telephone functions accessed by telephone function keys 60 may include mute, speaker, line select, conference, hold, transfer, volume control, and the like.

However, the transaction entry device 12 is further characterized by display 20 with touch screen 64, mode switch/computer function keys 66, optional retractable keyboard 68, and optional magnetic/smart card reader 70, which facilitate operation of the transaction entry device 12 in the transaction entry mode. A memory card reader may also be accessed via a door (not shown) as in a laptop computer. Preferably, display 20 is a super twisted, high contrast, reflective liquid crystal display (LCD) with a minimum of 20 characters per line and 16 lines (preferably, 40 columns by 24 lines), while touch screen 64 is preferably a clear pressure sensitive keyboard made up of 224 keys (16 rows of 14 keys) attached to the face of the LCD. Preferably, the LCD is also available as a backlit unit. Of course, touch screen 64 is not necessary if optional keyboard 68 is provided. In addition, a battery backup 71 (FIG. 6) may also be provided; alternatively, the battery 71 may be the primary power source for a portable (cellular) embodiment of the transaction entry device 12 in accordance with the invention.

FIG. 5b illustrates several of the connections to transaction entry device 12. Typically, transaction entry device 12 includes a handset (headset) jack 72 for connecting optional handset (headset) 54 to telephone electronics 14 when it is desired to communicate more privately than when only microphone 62 and speaker 16 are used. A video input port 74 is also provided for connecting conventional data compression circuitry 75 within the transaction entry device 12 (FIG. 6) to an optional video camera which provides picture phone type video or to a facsimile device or scanner. Such video data may be appended a frame at a time to the end of a data transaction in miscellaneous processing field 50 to create a multimedia data transaction as described above with respect to FIG. 2. A video output port 76 is also provided for providing decompressed video or facsimile data from data decompression circuit 77 (FIG. 6) to a video receiver, a high quality computer monitor, a facsimile device, and the like. Such data may also be provided to printer port 82 or 84 as desired. A multi-line phone jack (modem interface) 78 is also provided. Preferably, modem interface 78 provides separate modem connections for the telephone electronics 14 and the data transaction assembler 18, although only a single modem connection is necessary.

An optional infrared transceiver 80 is further provided for enabling remote control operation of television and stereo equipment and the like in response to data transactions transmitted/received by the transaction entry device 12. Infrared transceiver 80 includes an internal signal generator chip which reads parameters stored in data transaction assembler 18 for determining the appropriate transmission frequencies for the infrared signals. Control of the infrared devices is then provided through menus on the display 20. Additional infrared transceivers 80 may also be provided on each corner of the housing 52 so that the infrared signal will cover more area (each transmitter typically covers about 600 circumference). All such devices are known to those skilled in the art and thus will not be described in detail here.

A computer interface (RS-232) serial port 82 and parallel port 84 is also provided for transmitting/receiving data to/from another computer device and for providing output to a printer. A power input port 86 and a keyboard input 88 are also provided. Keyboard input 88 accepts a connection from a standard keyboard or a folding type keyboard (not shown) which may be used in addition to, or in place of, retractable keyboard 68. An optional removable PCMCIA memory card interface 89 (FIG. 6) for updating the operating instructions of the data transaction assembler 18 and an optional RF transceiver 90 (FIG. 6) for wireless networking to other electronic equipment may be provided on the transaction entry device 12 as desired.

FIG. 6 is a schematic diagram of the electronics of the transaction entry device illustrated in FIGS. 5(a) and 5(b). Corresponding reference numerals for corresponding elements are used in FIGS. 5(a), 5(b) and 6. As shown in FIG. 6, in addition to the elements described above with respect to FIGS. 5(a) and 5(b), the transaction entry device 12 may include a simple voice recognition circuit 91 which permits voice selection of menu options and the like. In "voice selection" mode, the user would voice "1", "2" or "3" depending on the desired menu selection, and the voice would be picked up by microphone 62 on the housing 52 of the transaction entry device 12 and recognized by voice recognition circuitry 91. The proper selection signal would then be sent to the data transaction assembler 18. Similarly, the data transaction assembler 18 may provide audible output using a conventional voice synthesizer 92, which provides the audio output to the user via speaker 16 and to a caller via modem interface/telephone line connection 78. The voice synthesizer 92 may, for example, allow certain data transactions to be audibilized for a blind person who cannot make selections from a conventional video display. In addition, a voice recorder 93 may also be provided to record portions of telephone calls, portions of voiced data transactions, or a caller's message as when using a conventional digital answering machine. On the other hand, voice recorder 93 may be provided in database server 28 for use in storing/forwarding audible messages to the database 30.

As noted above, the transaction entry device 12 is characterized by data transaction assembler 18, which controls the creation of data transactions in the transaction entry mode. As shown in FIG. 6, data transaction assembler 18 is implemented in hardware using a conventional microprocessor 94, such as an Intel 80386SX (20 MHz or higher) or equivalent, a TAS PROM 95, a form/menu memory 96, and a transaction buffer (RAM) 97. In a preferred embodiment, TAS PROM 95 is a flash PROM which holds 1 MB of control data (firmware) for the microprocessor 94 (such as the microcode for the algorithms of FIGS. 7–10 below), while form/menu memory 96 is a flash memory which holds 1 MB of data transaction menus and forms. Transaction

5,805,676

13	14

buffer 97, on the other hand, only needs to be as large as the largest data transaction, and may hold, for example, up to 128 kB of transaction data including application and operating system variables. Preferably, TAS PROM 95 and form/menu memory 96 are updated by downloading data streams containing new instructions and/or forms and menus over a conventional data bus 98 via modem 78, magnetic card interface 70, or via a removable memory card read by memory card interface 89 as necessary. Alternatively, additional flash memory elements may be added as additional applications are added to transaction entry device 12. Transaction buffer 97 may also be expanded to handle transactions of any size or type, including multimedia applications in which video and/or audio data is appended to data transactions.

Those skilled in the art will appreciate that the transaction entry device 12 may be docked into a docking station of a network. RF transceiver 90 may be used for wireless communications in such an environment. In addition, those skilled in the art will appreciate that the transaction entry device 12 may be implemented as a battery operated portable device which is a cross between a laptop computer and a cellular telephone of the type illustrated by Paajanen et al. in U.S. Pat. No. 5,189,632, for example. In such an embodiment, an optional headpiece could be provided, as well as a microphone and speaker arrangement in the fliptop. Of course, the liquid crystal display screen 20 would typically be reduced in size to, for example, 40 columns by 12 rows, and the touch screen 64 may be eliminated. However, most of the other options of the embodiment of FIGS. 5a and 5b would preferably remain so that the portable unit could also be used at a desk as desired. The electronics of the transaction entry device 12 would otherwise be as illustrated in FIG. 6 except for certain size and shape considerations well within the skill of those skilled in the art.

2. Software

As will be apparent from the following description, data transaction assembler 18 does not utilize a conventional operating system to control the operation of microprocessor 94. On the contrary, TAS PROM 95 stores simple firmware algorithms (FIGS. 7–10) operating in a kernel fashion for navigating a user through menus and forms provided from form/menu memory 96 for particular applications, and it is the resulting data streams which control the microprocessor 94 at any point in time. In other words, the data streams from the TAS PROM 95 and the data streams from the form/menu memory 96 together reconfigure microprocessor 94 into a special purpose processor for each application specified by the forms. The microcode of the TAS PROM 95 and the parameter streams from the form/menu memory 96 thus operate together as a simple form driven operating system for microprocessor 94 for all applications and is the sole code used to control microprocessor 94 (i.e., no conventional operating system or application programs are provided). As a result, the microprocessor 94 may be reconfigured into a new special purpose computer with each new form read from form/menu memory 96, and such forms/ menus may be added at any time by reading in the appropriate data streams from a memory card or from an external database server 28 or by adding an additional PROM. A specific implementation of the TAS firmware stored in TAS PROM 95 will be described below with respect to FIGS. 7–10.

Thus, the TAS PROM 95 contains control data (firmware) for the microprocessor 94 and resides in each transaction entry device 12 for generating a template for a data trans-

action from a data stream stored in form/menu memory 96 (or received directly from a memory card or external database server) and from data input by a user or retrieved from an external database or magnetic card, smart card, and the like. The TAS firmware and the selected template together control the behavior of the microprocessor 94 by logically defining a table of menu options and/or database interfaces which are navigated through by the user. As noted above, the user navigates through a series of menu selections by selecting another menu, a form, or a process. Once the data transaction for a desired application is completed, it is transmitted out for "explosion" into all of its component parts for storage. In this form, the TAS firmware from TAS PROM 95 and menus and forms from form/menu memory 96 of the invention together replace a conventional operating system and individual application programs. Indeed, the invention permits the transaction entry device 12 to be completely operating system independent.

The data transaction assembler 18 of the invention is connected via a predetermined protocol stored as instructions within TAS PROM 95 to a database server 28 and its associated database 30. As noted above, the database server 28 associated with a particular transaction entry device 12 operates as a repository of the data transactions created by the transaction entry device 12 and as a supplier of data to the transaction entry device 12 for completing the forms and providing additional forms, menus, processes, and the like. Since the system of the invention is operating system independent, there are no hardware or software limitations on the characteristics of the database server 28.

The parameter set making up the individual forms are typically provided by database server 28 as a stream of data via modem and stored in form/menu memory 96, while any downloaded instructions are stored in TAS PROM 95. Linkage between data transaction assembler 18 and its database server 28 is preferably provided via a dictionary program specific to each database server 28. This dictionary program knows the characteristics of each field of each form for each data transaction and is used by the database server 28 to "explode" the received data transactions into their component parts.

Preferably, at power on, data transaction assembler 18 automatically prompts the user with a "Download Parameter Streams" command so that the user can load into form/menu memory (flash memory) 96 from an external source the desired streams of menu and form data for the desired application. The "download parameter" process will then be initiated by dialing the external database server 28 initiating the connection via the modem interface 78. Once connected, the transaction controller 36 of database server 28 will transmit the requested parameter stream. The data transaction assembler 18 will load the received data stream into form/menu memory 96, and, upon completion, the prompt "Executive Menu Ready" will be presented on the display screen 20. The executive menu then will be automatically presented to the user for selection of the desired menu, form, or process.

Upon initiation of the transaction entry mode by the user, data transaction assembler 18 calls a set of panel parameters from form/menu memory 96 and paints a form onto display screen 20. These forms are either menus for navigating to particular forms or a form into which data is entered by the user. As will be explained below, the menus provide functionality through simple menu selection. The form on the display screen 20 is completed by the user by entering the appropriate data using touch screen 64 or optional keyboard 68. Alternatively, the requested data may be read in from a

5,805,676

15

memory card via memory card interface 90, from a magnetic strip on a swipe card or smart card via magnetic card interface 70 or memory card interface 89, or voice input via voice recognition circuit 91. In addition, a request may be sent to the database server 28 associated with the transaction entry device 12 for data needed to populate certain fields in the present form. The type of data entry is requested from a subset of options presented to the user upon pressing a "?" key or a "Request for More Information" button. This request will give the user several options to choose from, such as data entry using keyboard 68, touch screen 64, swipe card via magnetic card interface 70, memory card via memory card interface 89, by voice annunciation of the number of the item in the menu via voice recognition circuit 91, or via modem from a database 30. Hence, the data transaction created by the data transaction assembler 18 may or may not make use of stored data for reducing the amount of data entry required of the user.

When a data entry option is selected, data transaction assembler 18 does one of the following: another set of parameters is called up and another form is painted, the correctness of the selection is verified and a set of options for selections is presented based on interactions with stored data, the completed data transaction is transferred via modem to database server 28 for storage in database 30, or data values are requested from database 30 for incorporation within the transaction buffer 97. In a preferred embodiment, selections from the menu are made by touching the appropriate place on the menu using touch screen 64; by voice annunciation of the number of the menu item via microphone 62 and voice recognition circuit 91; by using one of the computer function keys 66 to run a cursor up the menu, another key to run the cursor down the menu, and a third key to make a selection in a conventional manner; or by using keyboard 68 as a selection device. When the keyboard 68 is used, the keyboard keys may be used to control a cursor, with the "enter" key being used for making a selection; alternatively, the number of the item selected may be entered and the "enter" key pressed to make the selection. Once the selection is made, the appropriate form is extracted from form/menu memory 96 as a stream of data.

Alternatively, in addition to presenting a menu for selection or completing a form, the data transaction assembler 18 can also present a menu selection for initiating a process such as calculation of an interest rate using one or more fields in the form, the finding of a mean, the finding of a name, or searching for entries for a particular date. These processes may be stored in TAS PROM 95, form/menu memory 96, in an off-line server where they are initiated, or any other place where they may be loaded down to the operating portion of the transaction entry device 12. In a preferred embodiment, processes are generally initiated in the database server 28 by sending a data request to the database server 28, processing the data in the database server 28, and then returning the answer as a data stream or report back to the transaction entry device 12.

A process typically initiates a data string which calls a process on an external machine. For example, the transaction entry device 12 may be used to download and store control signals for infrared control of various devices using infrared transceiver 80 (FIGS. 5 and 6). The form of the control signals will depend upon the signal storage in an optional infrared chip, which can be loaded by the data transaction assembler 18 or by an off-line device via modem or through the air using RF transceiver 90 for direct digital transfer in wireless form. In addition, in the case where the transaction entry device 12 is used in a medical office, for

16

example, the process may be used to transmit a prescription to a pharmacy or mail order house using prestored modem numbers or may enable the physician to call up a list of phone calls to make for the day or a list of the followup appointments for a particular date. In other words, the TAS firmware can also "explode" the data transaction into all of its ancillary parts for transmission to numerous records in one or more databases.

A preferred embodiment of the TAS firmware will now be described with respect to FIGS. 7–10.

As noted above, the transaction assembly (application) server (TAS) is a data stream stored in TAS PROM 95 which together with the forms from form/menu memory 96 create a simple form driven operating system which provides the necessary control data (firmware) to microprocessor 94 so that no conventional operating system is necessary. FIG. 7 is a flow diagram of a menu driven TAS in accordance with a preferred embodiment of the invention. As illustrated, the TAS firmware starts at step 100 and fetches an initial menu from form/menu memory 96 at step 102. The initial menu is prompted within a few seconds of booting the TAS firmware after the system logo. The initial menu typically presents the options of downloading a parameter stream from the database server 28 for enabling additional functions or printing an executive menu. If the executive menu is selected, the executive menu is retrieved from form/menu memory 96. The executive menu contains numerous application options to the user, namely, selection of a form, another menu, a process, or an automatic switch to telephone mode, one of which is selected at step 104. The data streams in form/menu memory 96 may be distinguished as to type (form, menu, or process) by appending a code such as "F" for form, "M" for menu, and "P" for process, and as to number by appending a form, menu, or process number at the beginning of the following data stream. These codes are recognized by the TAS firmware, and it acts accordingly.

If the option selected at step 104 is a form, the proper form (data stream for form F_{xy}) is fetched from form/menu memory 96 at step 106, a transaction buffer 97 equal in length to the size of the record associated with the form F_{xy} is formed in RAM, the form is stored in the transaction buffer 97, and a connection is made to the appropriate database server(s) 28. The data stream for the selected form will consist of prompts, print locations for the prompt, data entry points, print locations for the data entry start, data entry length, and a code as to the nature of the data entry. This code can be numeric, alphanumeric, a cross-reference to stored data or previously entered data, a formula for the creation internally to data transaction assembler 18 of the result from previously entered data, or an external request for data, help, or reformulated values.

The data stream entered into the fields of the form will not only indicate the location for the printing of the prompt and the field for data entry, but also the size of the field and the storage, a start point within the transaction buffer for the stored element, and the type of data: alphanumeric, numeric (floating point or integer), date, and the like.

If it is determined at step 108 that the requested form is actually a menu (M_{xy}), a hidden set of codes pointing to the form F_{xy} that the selection will lead to is read, and control branches back to step 104 for selection of another menu or form. When a menu is chosen, each item has its sequential number, its descriptor, and a code for what it will "call" (another menu, form, or process). In other words, each choice has associated with it a series of item codes which branch out to another form, menu, or series of tests upon the

17

data entered. A menu also has a numeric code for each of the storage areas and a special code including a security code for certain menu items, process codes of forms within the menu, or a pointer to the process code. A pointer may also be provided in the menu for processes to be performed off-line (i.e., in an associated database server **28**).

If a process (P_{ry}) is selected at step **104**, the database server **28** is notified that something is requested from its database **30** or that some processing of data is requested. For example, the data transaction assembler **18** may send a user "?" inquiry to the database server **28** so that options may be returned to the data transaction assembler **18** for presentation to the user for selection. The process triggers an external process of database server **30** with a parameter stream, and control is either returned to the data transaction assembler **18** or control is held up until the process is complete, in which case a message is sent back to the data transaction assembler **18**. This message can be a report, selected data, a value resulting from a calculation, and the like. Processing such as checking detectors and the like may also be performed locally by data transaction assembler **18**.

Once the desired form is selected for the user's application, the form is processed at step **110** in accordance with the steps outlined in FIGS. **8–10**. As an entry is made in each field, it is automatically stored within the input buffer area of the transaction buffer **97** at its assigned location and in the dictated format. At any time, the entire form may be exited with automatic return to the menu which called it or the form can be cleared for data reentry. Once the form has been processed and transmitted to the appropriate database server(s) **28**, the database server connection is terminated and the user is presented at step **112** with the last menu from which the user made his or her selection. Alternatively, the executive menu will be called up as a default menu.

If the user indicates at step **114** that he or she wishes to continue to complete a new form, control branches back to step **104** for menu selection and a new database server connection is made as appropriate. This process is repeated for each form. When no further selections are desired, the TAS firmware is exited at step **116**.

FIG. **8** is a flow diagram illustrating a technique for processing a form (step **110**) to create a data transaction in accordance with the invention. As illustrated, the process of FIG. **8** starts at step **118** and initializes a transaction buffer **97** at step **120** for storage of the data transaction as it is being created. In other words, if there is a form for the requested application, it is moved from form/menu memory **96** to the transaction buffer **97**. If the requested form is not present in form/menu memory **96**, an error message may be sent or a request may be sent to database server **28** for a download of a data stream containing the parameters for the requested form. Preferably, transaction buffer **97** is at least as large as the largest data transaction and serves as an assembly area for the data transaction. Preferably, read and write buffers are formed so that transmit and receive buffers to/from modem interface **78** are available. Of course, transaction buffer **97** may be made larger for this purpose.

Once the transaction buffer **97** is initialized at step **120**, the display screen **20** is cleared and the selected form is initialized to its first page at step **122**. The first page is then presented to the display screen **20** at step **124**. At step **126**, the user completes the form page on a field by field basis using any of the data entry techniques described above and the field controls of FIGS. **9** and **10**.

The transaction buffer **97** collects the data associated with the form presented to the user on display screen **20** and

18

contains appropriate locations for each separate data element. Upon completion of the data transaction, the contents of the transaction buffer **97** are transferred to the appropriate database server(s) **28** via modem or via wireless, preceded by a set of codes (field **44**, FIG. **2**) which identify the type of data transaction and followed by a string of process identifiers for the database server(s) **28** to use in its programs in creating additional transactions and in storing the data and all ancillary data transactions in the regular file format of the database **30** associated with the database server(s) **28**. As a result, the data transaction created in the transaction buffer **97** has a one-to-many relationship to the data stored in the database **30**.

If the user decides to abort the processing of a form at any time (step **128**), the form processing routine is exited at step **129**. Otherwise, it is determined at step **130** whether the user wishes to go back a page (for a multi-page form) to correct a data entry. If so, control returns to step **124** for presentation of the earlier page. If the user does not wish to examine or edit a previous page, it is determined at step **132** whether the current form has another page which has not been displayed for completion by the user. If the form has more pages, the routine moves to the next page at step **134**, and it is determined at step **136** whether the move to the next page was successful. If so, control returns to step **124** for presentation of the next page. Of course, the process of calling a subsequent page in a form or another form upon completion of a form can be dependent upon an automatic call of that page or form sequence or the ability to jump sequence (i.e., skip pages) depending upon a value in any one field that has been entered. In any event, if there are no more pages in the form or if the move to the next page was not successful, the end of the form is marked with a code and the transaction is saved at step **138** by sending the data transaction to the appropriate database server(s) **28** for storage in its associated database **30** and "explosion" for storage of data in other databases **40**. If it is determined at step **140** that the save was not successful because of a modem error and the like, control returns to step **122** and the process is repeated. If the data transaction was successfully saved, the form processing routine is exited at step **129** and the last menu used is presented (step **112**).

Optionally, stored procedures within any data transaction form (field **50**, FIG. **2**) are executed at the appropriate time within the flow of the form processing routine before it is exited. However, these processes may be deferred and performed by the database server **28** if needed.

FIGS. **9**(*a*) and **9**(*b*) together illustrate a flow diagram of a technique for completing and editing the fields of a form (step **126** of FIG. **8**). The field completion routine starts at step **142** and first determines at step **144** whether an abort or a valid page move request is pending. If so, the field completion routine is exited at step **146**. However, if no abort or page move request is pending, the field data for the first field of the transaction buffer **97** is entered at step **148**. As noted above, this field data may be entered via keyboard **68** or touch screen **64**, swiped in via magnetic card interface **70**, read in from a memory card via memory card interface **89**, read in via modem interface **78** from database server **28**, or designated by voice entry. Preedit processing of the field data is then performed at step **150**. Such pre-edit processing may include, for example, setting default values, performing calculations, establishing links to data in other files, looking up and writing data to files already linked to the present form, spawning another form, performing special updates of the display screen **20**, hiding fields from view by the user, and the like. Such pre-edit processing may also be used to

5,805,676

19

determine whether modifications or actions in the present field may invalidate an entry in another interrelated field. If so, appropriate measures are taken to update all affected fields or to prevent such problems by setting appropriate default values.

The field completion routine then checks for field errors at step 152 on the basis of the default values and the like set at step 150. If there is no field error at step 152, it is determined at step 154 whether the operator will be permitted to edit the field in the absence of a field error. If so, or if a field error was found at step 152, the operator edits the field at step 156. If the operator editing is bypassed, control proceeds directly to post-edit processing at step 158, which performs essentially the same functions as pre-edit processing step 150 except that the data may be specially validated. The field is then checked yet again at step 160 for a field error. If a field error is found at step 160, control returns to step 144 for processing the next field or exiting, as appropriate.

If no field error is found at step 160, it is determined at step 162 whether the generic field validation routine of step 164 (FIG. 10) is to be skipped. If so, control proceeds to step 166, where the field is once again checked for a field error. However, if generic field validations are desired, control passes to the routine of step 164 (FIG. 10). If no field error is found at step 166, the field is saved to the transaction buffer 97 at step 168 and the updated field value is painted on the display screen 20 at step 170. If the user then desires to check a previous field at step 172, control passes to a previous field at step 174 and the field completion routine is repeated for the previous field. However, if no previous field is to be checked and if it is determined at step 176 that a further field is present, control passes to the next field at step 178 and the field completion routine is repeated for the next field. This process repeats until the last field is completed and the routine exits at step 180. Control then returns to FIG. 8 for processing a different page of the form.

Each form may be processed in one or more modes. In the input mode, described above, the data transaction is created and transmitted to the database server 28. However, in edit mode, upon entering the ID of a particular record, that record is read from an external database 30 or 40 into transaction buffer 97 for editing. Preferably, a record of the edits is maintained to provide an audit trail. In view mode, upon entering the ID of a particular record, that record is similarly read from an external database 30 or 40 into transaction buffer 97 but for display only. Finally, in delete mode, an entire record can be deleted from the database 30 or 40 if the user has proper security clearance.

FIG. 10 illustrates how the TAS firmware validates the fields of each data transaction. As shown, the field validation routine starts at step 182 and first determines at step 184 what field type is present. If the present field is an alphanumeric field, control passes to step 186 where the field defaults are processed. It is then determined at step 188 whether the user knows the values allowed for this field. If not, and data is to be implanted in that field, an implant table is searched at step 190. A "?" may be used by the operator to indicated that he or she does not know the values allowed for this field and wishes to search the implant table. A list of possible values can then called up that match the data entered thus far. From this list, the operator can scroll the list and select the value that will complete the data entry. However, if the value is not found in the list, a field error is generated at step 192 and the field validation routine is exited at step 194. If the value is found in the list, control passes to step 200.

20

On the other hand, if at step 188 it is determined that data need not be added (implanted) into the present field, control skips to step 196, where it is determined whether the present field type is a field which sets up an event in which the present field (along with its form) can be linked to any record of any file or files (one to many) of any database for the purpose of data verification and/or data extraction. If so, control passes to step 198, where the data from the present field along with any other data previously gathered is used to make the desired link. As in the data implant step 188 noted above, the user may enter a "?" to get the information needed to make this link. If the data for the link is not found, a field error is issued at step 192 and the field validation routine is exited at step 194. However, if the data for the link is found, the field is checked for blanks at step 200 and a field error is issued at step 192 if blanks are present in the field but are not allowed. If no blanks are found in the present field, or if blanks are found but are allowed, the field validation routine is exited at step 202.

If it is determined at step 184 that the present field is a numeric field, the field is checked at step 204 to determine if the character set is valid. If so, the precision of the numbers is adjusted at step 206, as necessary, and the range and scope of the numbers are checked at step 208 to make sure the field entries satisfy the boundary conditions (e.g., no dividing by zero). If the character set is not valid at step 204 or the range and scope of the numerals is not valid at step 208, a field error is issued at step 210 and the data validation routine is exited at step 212. Otherwise, the field validation routine is exited at step 214.

If it is determined at step 184 that the present field is a date/time field, the field is checked at step 216 to determine if the character set is valid. If not, a field error is issued at step 210 and the field validation routine is exited at step 212. Otherwise, a routine of the TAS firmware checks the date/time entry at step 218 to determine if it has the correct format by performing range checking and the like. If the date/time entry does not have the correct format, a field error is issued at step 210 and the field validation routine is exited at step 212. Otherwise, it is determined at step 220 whether the present field contains a date. If not, the data validation routine is exited at step 221. If so, the date is checked at step 222 to see if it contains a weekend, and, if so, checks at step 224 whether a weekend date is an acceptable reply for this field. It is then determined at step 226 whether the calendar file is to be checked, and if so, the calendar file is checked at step 228 to see if the date is valid (e.g., not a February 30 and the like). Finally, it is determined at step 230 whether a warning date has been exceeded, and if so, a field error is issued at step 210 before the field validation routine is exited at step 212. Otherwise, the field validation routine is exited at step 221.

Those skilled in the art will appreciate that, in order to maintain security, the TAS firmware may also present a security form for password entry to the user. The security form and ID of the transaction entry device 12 is then encrypted and transmitted to the database server 28 associated with the particular data transaction assembler 18. Transaction controller 36 of that database server 28 will then act as the transaction controller for that data transaction assembler 18 and will check passwords and the like during operation to make certain that data security is not breached. Database servers 28 may disable a data transaction assembler 18 if unauthorized use is attempted. In this manner, only the appropriate person may view each menu. Of course, a different number of security levels and different executive menus may be presented as desired, all under control of the transaction controller 36.

5,805,676

| 21 | 22 |

C. Database Server 28

As noted above, the database server **28** acts as a vehicle for separating data transactions created by the data transaction assembler **18** into the component parts thereof which may be stored directly in one or more databases **30** and **40**. In other words, the database server **28** explodes the initial data transaction into data transactions for many different files for updating records in the files, and the like. Also, the database server **28** may be virtual as well as real, exist in a single machine or in multiple machines, in whole or in part.

Generally, the database server **28** handles any and all data transactions received, manipulates data in the data transactions, spawns or starts processes or reports requested by a data transaction, and explodes the received data transactions into all sorts of data transactions that were spawned by the initial data transaction. Database server **28** can also update values in existing records and can switch to a process for processing values in the records as necessary. In this manner, a single data transaction can define actions causing multiple files to be updated. Database server **30** also handles requests from the data transaction assembler **18** and processes them as needed. Such requests may include data I/O requests, data locking and unlocking, report processes, and requests for new forms or menus. Those skilled in the art will appreciate that database server **28** maintains the one-to-many relationships that exist between the user and the system of the invention, the one-to-many presentations to the user and files in the databases **30** and **40**, and the one-to-many data transactions and the ancillary records, updates, and postings as may be required to diverse computer files of numerous databases **40**, the transaction entry device **12** and the database servers **28**.

As noted above, transaction buffer **97** collects the transaction data associated with the form presented to the user via display screen **20**. The transaction buffer **97** is the image of the data transaction with appropriate locations for each separate data element. The contents of the transaction buffer **97** are transferred to the database server **28** via modem interface **78** or via RF transceiver **90**, preceded by a set of codes **44** (FIG. 2) which identify the type of transaction followed by a string of process identifiers for the database server **28** to use in its programs, in creating additional data transactions, and in storing the data and all ancillary transactions within the database **30** in the regular file format of the database **30**. In other words, the database server **28** determines what type of action to take based on the type of data transaction received, "explodes" a data transaction into a plurality of other data transactions for transmission to other databases, as appropriate, and converts the data for its associated database **30** into the proper file format. Of course, each database server **28** is different from each other database server **28** since it will handle different types of data transactions, have different operating system characteristics, and different file conversions to make in accordance with the file formats of its associated database **30**. For example, the database server **28** may operate under an operating system such as Unix, Windows, or DOS, where the operating system provides the database server **28** with links to the hardware functions normally handled by an operating system. Preferably, the database server **28** also operates with menus, forms, and the like in the same fashion as the data transaction assembler **18** except that it stores the data transactions in its associated database **30** as transaction files.

As just noted, the purpose of the database server **28** is to process the data transaction from the data transaction assembler **18** and to either explode the data transaction into all of its related components for storage, to handle the storage of

items from the explosion process, to store the data transaction itself for reference purposes, and to act as a supplier of information to the data transaction assembler **18** in response to requests during the creation of the data transaction and during the downloading of parameters for menus and forms to the data transaction assembler **18**. If desired, the database server **28** can also supply information back to the data transaction assembler **18** after a data transaction is received or can initiate a process leading to the delivery of a report, data, or menu to the data transaction assembler **18**. In addition, the database server **28** and data transaction assembler **18** can reside on the same machine so long as the database server's operating system recognizes the TAS firmware or the TAS firmware is modified for use with the operating system of the database server **28**.

D. Applications of the Invention

As outlined above, the present invention includes a point of transaction device which presents a menu to a user from which an option is selected. A form tailored to the selected option appears for guiding the user through data entry. The full details of the data transaction are captured as data is entered by the user. Modem interaction with a central database(s) or a user database(s) allows for interaction with help and verification of certain entered data. The completed transaction is then transmitted to the central or user database for further processing and storage. Data input can also be provided via a swipe card or smart card, from data received from any database accessible via the modem interface, or other known methods.

A data transaction system of this type may be used for many applications. For example, in a first, presently preferred, application, the transaction entry device **12** is located in a medical office for entry of patient data. In this application, a swipe card identifies the patient, a smart card identifies the doctor, and the modem connection allows the entire claim transaction to be entered and transmitted to the insurance companies for processing. The patient records may also be automatically updated and prescriptions created, given to the patient, transmitted to the pharmacist, and transmitted to the payor and patient record. Patient instructions such as special diets, exercises, treatments, appointments, and the like may be printed from the data transaction form at the doctor's central computer. In addition, a video image or picture provided via video input **74** and compressed by data compression circuitry **75** permits an image of a medical condition such as a rash to be appended to the data transaction (in miscellaneous processing field **50** of FIG. 2) for transmission with the patient's name, the date, a description of patient symptoms, and the like. Similarly, a recorded heartbeat may be appended to the end of the data transaction for transmission with the patient data.

The data transaction entry system of the invention also has numerous home uses. In a preferred home use, the transaction entry device is used for performing bank transactions from the home. In this case, forms would be made available by the bank for different types of bank transactions. These forms would then be downloaded to the transaction entry device in the customer's home and used in creating and transmitting data transactions to the bank computer for off-line processing.

As another example, the user may dial-up a 900 number to get an interface to a central database which will download codes into TAS PROM **95** or form/menu memory **96** which enable the generation of infrared signals at certain frequencies. The user needs only to specify the type, make and

5,805,676

23

model of any electronic device to be controlled in order to get the desired code. Then, to operate any electronic device in the home, the user would be directed by menu prompts. The transaction entry device 12 would then emit an infrared signal via infrared transceivers 80 to operate the electronic device, initiate a call via modem for a broadcast program, or initiate timed requests for video recording, turning the video recorder on and off, and the like.

For other home uses, the transaction entry device 12 may also initiate, via menu prompts, sequences for turning on and off various household devices including alarm systems, coffeemakers, and the like. In this mode, the transaction entry device 12 may receive an RF or infrared signal indicating that a burglar or fire alarm has been activated and call up a form for calling the police or fire department, as appropriate. A call to the transaction entry device 12 may then be used to turn off the burglar or fire alarm by changing a field in a form which instructs the infrared transceiver 80 or RF transceiver 90 to send an appropriate control signal to the alarm device. This feature may also be prompted from a car phone via remote initiation of the form performing this function.

The transaction entry device 12 may also control all household telephone use as well as controlling the answering machine and keeping a telephone transaction log. The user may also pay household bills by completing an appropriate form and transmitting the form to a payee such as a credit card company, a bank, and the like. In short, the transaction entry device will permit the owner to connect to a remote database without owning a conventional computer system with an operating system and the like.

For personal applications, the transaction entry device 12 may be used to initiate a facsimile transmission, to provide telephone lists with automatic dialing upon selection, to provide expense accounts, personal scheduling, tax record keeping, and the like, and to provide direct access to travel information. For example, the database server 28 may be an airline reservations system. In this application, the data transaction assembler 18 dials the modem of the airline reservations system when the user requests data entry into an airline reservations form available at the user's transaction entry device 12. The data transaction device 18 modems the database server 28, and the operating system of the database server 28 selects interface programs for the airline reservations system. The interface programs call the database servers 38 of the airlines, retrieve the appropriate menu from database 40, and modem the menu to the data transaction assembler 18. The data transaction assembler 18 then displays the airline reservations menu on its display screen 20 for completion and transmission back to the airline reservations database server for processing. The swipe card may be used to provide credit card payment information and may be updated by permitting the data transaction assembler 18 to write to the swipe card. The user may also access frequent flyer club and mileage data, special offers on hotels, cruises and other travel, and the like.

In another home (or business) use, the transaction entry device 12 may be used to eliminate conventional phone mail greetings by enabling the caller's transaction entry device 12 to read in a set of visible menus from the called party's voice mail menu so that the calling party may select the desired options using a visible menu rather than a voiced menu. In other words, the caller would not have to wait through the litany of voiced phone mail options before making a selection and could make the desired selection right off of his or her own display. This would be accomplished by selecting a process from the menu of the transaction entry device 12

24

which will create a "visible" menu. When such a process is selected, the telephone electronics 14 or modem interface 78 makes a telephone connection to a remote phone mail system. Once the connection is made, the data transaction assembler 18 sends a data request for a visual representation of the phone mail menu of the remote phone mail system via the telephone connection to the remote phone mail system. A data stream containing the visual representation of the phone mail menu from the remote phone mail system is then returned via the telephone connection and stored in form/menu memory 96 and presented to display screen 20 of the transaction entry device 12 for selection using the techniques described herein. When menu items are selected from the "visible" voice mail menu, the data transaction assembler 18 creates a data transaction indicating which menu item was selected and sends the data transaction to the remote phone mail system via the telephone connection. Based on the menu selection, the remote phone mail system then returns a data stream containing a visual representation of the next phone mail menu via the telephone connection for storage in form/menu memory 96 and display on display screen 20. This process is repeated until the calling party is required to leave a message or the called party is reached. Such a system would be particularly helpful for interacting with voice mail systems, such as those at government offices, where numerous options are presented for selection.

Those skilled in the art will appreciate that the invention is unique by virtue of its ability to generalize applications to forms so that no code need to be written to implement a particular function. However, if code is needed or if multimedia data is to be part of a data transaction, it can be attached to a form which is stored as a parameter stream in a stream of data. Also, though the transaction entry device 12 has been described as a computer workstation, it can also be used in conjunction with an optional off-line storage device as a self-contained workstation and database unit independent of traditional operating systems. The transaction entry device 12 can also be used with an additional optional plug in as a network server or as a user interface in a network docking station.

Those skilled in the art will also appreciate that the foregoing has set forth the presently preferred embodiments of the invention but that numerous alternative embodiments are possible without departing from the novel teachings and advantages of the invention. Accordingly, all such modifications are intended to be included within the scope of the appended claims.

I claim:

1. A system for entering transaction data into a remote database, comprising:

a data input device;

a display;

a data transaction terminal including a microprocessor, a form memory which stores a plurality of menus and forms for presentation to a user, and a form driven operating system which controls a process implemented by said microprocessor to present to said display for each process at least one form stored in said form memory as data streams, said at least one form being selected by said user from one of said menus using said data input device, said one menu providing said user with an option of selecting at least one of said at least one form, another menu, and an updating process, each form eliciting data input of a desired transaction type into said data input device by said user and including at least one prompt customized to said

264

5,805,676

25

desired transaction type, wherein said process implemented by said microprocessor is changed by changing said at least one form, and wherein when said user selects said updating process from said menu, data streams are downloaded to said form memory to update said menus and forms in accordance with said desired transaction type, said data transaction terminal further including means for formatting at least said data input by said user in response to said at least one prompt into a data transaction for transmission to said remote database; and

a database server associated with said remote database which receives said data transaction, creates from said data transaction, depending on said desired transaction type, at least one additional data transaction containing data for a particular record in said remote database, and stores said at least one additional data transaction in said particular record.

2. A system as in claim 1, further comprising a plurality of remote databases, wherein said database server further creates from said data transaction, depending on said desired transaction type, at least one ancillary data transaction containing data for a particular record in one of said plurality of remote databases besides said remote database and stores said at least one ancillary data transaction in said particular record.

3. A system as in claim 2, wherein said form driven operating system includes means for sending a data request to said database server, said database server accessing data corresponding to said data request in at least one of said remote databases and returning one of data responsive to said data request, a list of options for selection by said user, a value calculated from data contained in said data request, and a data report.

4. A system as in claim 1, wherein said form driven operating system comprises a transaction assembly server (TAS) which presents said data streams to said microprocessor for display on said display, and said formatting means comprises a transaction buffer which stores said data input into said data input device by said user in response to said at least one prompt until said data transaction is completed for transmission to said remote database.

5. A system as in claim 1, wherein said data transaction terminal further comprises a modem, a telephone and two telephone line connections, one for connecting said telephone to a telephone network, and one for providing a modem connection among said modem, said TAS, and said database server.

6. A system as in claim 5, wherein said data transaction terminal further comprises a mode switch for selecting a telephone mode in which said data transaction terminal operates exclusive of said TAS or a transaction entry mode in which said TAS operates exclusive of said telephone.

7. A system as in claim 4, wherein said data transaction terminal further comprises a modem, a telephone, a telephone line connection, and means for selectively connecting said telephone to a telephone network and said TAS to said database server via said telephone line connection.

8. A system as in claim 7, wherein said selectively connecting means comprises a mode switch for selecting a telephone mode in which said data transaction terminal operates exclusive of said TAS or a transaction entry mode in which said TAS operates exclusive of said telephone.

9. A system as in claim 1, wherein said database server comprises a modem, a data transaction queue for storing data transactions received from said data transaction terminal, and a transaction controller which processes the

26

received data transactions to extract physical relationships of data of said data transactions with records in said remote database.

10. A system as in claim 1, wherein said one menu further contains a remote process option, and when said user selects said remote process option from said one menu, data streams are downloaded via a modem to said form memory, said data streams containing control data for implementing functions designated by said selected remote process option.

11. A system as in claim 10, wherein said data transaction terminal further comprises an infrared transceiver and said control data comprises data for controlling a wavelength of energy emitted by said infrared transceiver.

12. A system as in claim 10, wherein said data transaction terminal further comprises a phone list memory for storing a phone list and said control data comprises data for updating said phone list.

13. A system for entering transaction data into a plurality of remote databases, comprising:

a data transaction terminal for capturing a data transaction having a one-to-many relationship to records of said plurality of remote databases, said data transaction terminal including a microprocessor, a form memory which stores a plurality of menus and forms for presentation to a user, and a form driven operating system which controls a process implemented by said microprocessor to present to said user for each process at least one form stored in said form memory as data streams, said at least one form eliciting data input of a desired transaction type into said data transaction terminal by said user, said data streams of said at least one form including at least one prompt customized to said desired transaction type, a format field which identifies said desired transaction type of said at least one form, a data entry field including said at least one prompt and spaces for said data input by said user in response to said at least one prompt, and a processing field for appending data particular to said desired transaction type, said data transaction terminal further including means for formatting at least said data input by said user in response to said at least one prompt into said data transaction for transmission to at least one of said plurality of remote databases;

a first set of database servers associated with a first set of remote databases of said plurality of remote databases, said first set of database servers receiving said data transaction, creating from said data transaction a plurality of ancillary data transactions having a one-to-one relationship to said records of said plurality of remote databases, and storing said ancillary data transactions in designated records of said first set of remote databases; and

a second set of database servers associated with a second set of remote databases of said plurality of remote databases, said second set of database servers receiving certain of said plurality of ancillary data transactions, and creating from said certain ancillary data transactions additional data transactions which are stored in application specific records of said second set of remote databases in accordance with said desired transaction type of said at least one form.

14. A system as in claim 13, wherein each database server of said first and second set of database servers comprises a modem, a data transaction queue for storing data transactions, and a transaction controller which processes a received data transaction to extract physical relationships of data of said data transactions with records of a remote database associated with said each database server.

15. A system as in claim 13, wherein said data particular to said desired transaction type includes at least one of audio and video data.

16. A data transaction terminal for providing data transactions to a remote database server which stores records in an associated database, comprising:

a data input device;

a display;

a telephone circuit;

a data transaction assembler including a microprocessor, a form memory which stores a plurality of menus and forms for presentation to a user, and a transaction assembly server (TAS) which controls a process implemented by said microprocessor to present to said display for each process at least one form stored in said form memory as data streams, said at least one form being selected by said user from one of said menus using said data input device, said one menu providing said user with an option of selecting at least one of said at least one form, another menu, and an updating process, each form eliciting data input of a desired transaction type into said data transaction assembler by said user and including at least one prompt customized to said desired transaction type, wherein said process implemented by said microprocessor is changed by changing said at least one form, and wherein when said user selects said updating process from said menu, data streams are downloaded to said form memory to update said menus and forms in accordance with said desired transaction type, and means for formatting at least said data input by said user in response to said at least one prompt into a data transaction for transmission to said remote database server; and

a mode switch for selectively connecting said telephone circuit to a telephone network in a telephone mode and said data transaction assembler to said remote database server in a data transaction entry mode.

17. A terminal as in claim 16, wherein said formatting means comprises a transaction buffer which stores said data input into said data input device by said user in response to said at least one prompt until said data transaction is completed for transmission to said remote database.

18. A terminal as in claim 16, further comprising a modem and two telephone line connections, one for connecting said telephone circuit to said telephone network, and one for providing a modem connection among said modem, said TAS, and said remote database server.

19. A terminal as in claim 16, further comprising a modem and a telephone line connection, said mode switch selectively connecting said telephone and said TAS to said telephone line connection.

20. A terminal as in claim 16, wherein said one menu further contains a remote process option, and when said user selects said remote process option from said one menu, data streams are downloaded via a modem to said form memory, said data streams containing control data for implementing functions designated by said selected remote process option.

21. A terminal as in claim 20, further comprising an infrared transceiver, said control data comprising data for controlling a wavelength of energy emitted by said infrared transceiver.

22. A terminal as in claim 20, further comprising a phone list memory for storing a phone list, said control data comprising data for updating said phone list.

23. A terminal as in claim 17, wherein said TAS presents one of said menus to said user for selection, said one menu

containing pointers to a plurality of forms, and upon selection of said at least one form from said menu by said user, said TAS initializes said transaction buffer and presents said at least one form to said display on a page by page basis for entry of said input data by said user.

24. A terminal as in claim 16, wherein said TAS processes said input data as it is entered in response to each prompt to determine if said input data satisfies predetermined conditions for input data entered in response to each said prompt.

25. A terminal as in claim 16, wherein said TAS sends a data request to said remote database server when said user requests assistance in replying to a prompt and inserts reply data from said remote database server into said data transaction in response to said prompt.

26. A terminal as in claim 16, wherein said TAS comprises means for creating from said data transaction, depending on said desired transaction type, at least one ancillary data transaction containing data for a particular record in said associated database and storing said at least one ancillary data transaction in said particular record in said associated database.

27. A terminal as in claim 26, wherein said ancillary data transaction creating means further creates from said data transaction, depending on said desired transaction type, an ancillary data transaction containing data for an application specific record in a secondary database and sends said ancillary data transaction to said secondary database for storage of said ancillary data transaction in said application specific record.

28. A terminal as in claim 16, wherein said data transaction assembler includes means for sending a data request to said remote database server, said remote database server accessing data corresponding to said request in said associated database and returning one of data responsive to said request, a list of options for selection by said user, a value calculated from data contained in said data request, and a data report.

29. A terminal as in claim 16, wherein said data input device comprises at least one of a touch screen associated with said display, a telephone numeric keypad, an alphanumeric keyboard, a memory card reader, and a magnetic card reader.

30. A terminal as in claim 29, wherein said alphanumeric keyboard comprises a retractable keyboard which retracts into a housing of said data transaction terminal.

31. A terminal as in claim 16, further comprising a video input terminal for receiving input video data and a video output terminal for providing output video data to a video monitor.

32. A terminal as in claim 31, further comprising a data compression circuit for compressing said input video data prior to including said input video data in a data transaction and a data decompression circuit for decompressing output video data prior to display on said video monitor.

33. A terminal as in claim 16, further comprising a computer I/O port for receiving input data from a computer device and providing output data to at least one of said computer device and a printer.

34. A terminal as in claim 16, further comprising a RF transceiver for providing a wireless connection between said data transaction terminal and a data processing device.

35. A terminal as in claim 16, further comprising a battery for providing power to said data transaction terminal for portable operation.

36. A terminal as in claim 16, wherein said data input device comprises a voice recognition circuit for accepting data input selections annunciated by said user.

5,805,676

| 29 | 30 |

37. A terminal as in claim **16**, further comprising a voice synthesizer responsive to said data transaction for audibilizing a portion of said data transaction to said user.

38. A terminal as in claim **37**, further comprising a voice recorder for recording at least one of said audibilized portion of said data transaction when in said data transaction entry mode and voice input from a called party when in said telephone mode.

39. A method of entering transaction data into a remote database using a data transaction terminal, comprising the steps of:

storing a plurality of menus and forms in a form memory of a form driven operating system of said data transaction terminal, each form including at least one prompt customized to a desired transaction type;

said form driven operating system controlling said data transaction terminal to accept input data of said desired transaction type using control data comprising at least one of said forms from said form memory;

a user selecting one of said menus using a data input device and said user selecting from said one menu at least one of said at least one form, another menu, and an updating process for further processing;

if said updating process is selected from said one menu by said user, downloading data streams to update said menus and forms in accordance with said desired transaction type; and

if said at least one form is selected from said one menu by said user, said form driven operating system presenting to a display a form for eliciting data input of said desired transaction type from said user, said user inputting data in response to said at least one prompt of said form using said data input device, and said form driven operating system formatting at least said input data from said user into a data transaction for transmission to said remote database and transmitting said data transaction to said remote database.

40. A method as in claim **39**, comprising the additional steps of:

receiving said data transaction at said remote database;

creating from said data transaction, depending on said desired transaction type, at least one additional data transaction containing data for a particular record in said remote database; and

storing said at least one additional data transaction in said particular record.

41. A method as in claim **40**, comprising the additional steps of:

creating from said data transaction, depending on said desired transaction type, at least one ancillary data transaction containing data for a particular record in an ancillary database different from said remote database; and

storing said at least one ancillary data transaction in said particular record in said ancillary database.

42. A method as in claim **39**, comprising the additional steps of:

sending a data request to a database server of said remote database; and

said database server accessing data corresponding to said data request in said remote database and returning one of data responsive to said data request, a list of options for selection by said user, a value calculated from data contained in said data request, and a data report.

43. A terminal as in claim **28**, wherein said remote database server is a remote phone mail system and said telephone circuit makes a telephone connection to said remote phone mail system, said data request being sent via said telephone connection and including a request for a visual representation of selection options of a phone mail menu of said remote phone mail system, and, in response to said data request, said remote phone mail system returning via said telephone connection a data stream containing said visual representation of said selection options of said phone mail menu, said visual representation of said selection options of said phone mail menu being presented to said display by said data transaction assembler for selection by said user using said data input device, and said data transaction assembler further sending data to said remote phone mail system via said telephone connection indicating which selection option was selected from said phone mail menu by said user.

44. A terminal as in claim **43**, wherein said remote phone mail system returns a data stream containing a visual representation of selection options of a next phone mail menu via said telephone connection in response to said data indicating which selection option was selected from said phone mail menu by said user.

* * * * *

Appendix VIII

Brief Biographical Summary of
Dr. Rocco Leonard Martino

Dr. Rocco L. Martino is the inventor of the CyberFone – the world's first smartphone – and the driving force behind the software systems permitting secure real-time video, voice and data linkages. Martino graduated Summa Cum Laude in Honors Mathematics and Finance from University College at the University of Toronto in 1951, went on to earn a Master's Degree in Physics in 1952, and a Doctorate in Aerospace Engineering from the University of Toronto Institute of Aerospace Studies in 1956. His discovery of the heating factors during the re-entry of space vehicles led to the development of heat shields that made space travel possible today. He is the Founder and Chairman of the Board of Martino Systems, Inc. and U.S. Robots, Inc., and was the Founder, Chairman and CEO of XRT, Inc., a global leader in providing complete treasury, cash and banking relationship management solutions for many of the world's largest corporations and government entities.

Prior to founding XRT, Inc., Dr. Martino directed the Aerospace Division of Adalia, Ltd, a firm headed by Sir Robert Watson Watt, the inventor of Radar; directed all activities in Canada for

UNIVAC, and worked with Admiral Grace Hopper on automatic programming systems; formed a partnership to create Mauchly Associates with Dr. John Mauchly, the co-inventor of computers, and spearheaded the Critical Path Method created by his company; served as the Chief Information Officer of the Olin Mathieson Corporation; and finally headed the Special Projects Group of Booz Allen and Hamilton.

Rocco Leonard Martino is also the author of five novels, twenty-seven nonfiction books, as well as scores of papers and numerous corporate monographs on computers, communications, networks and planning.

He served as Professor of Engineering and Chair of the Systems Engineering Department of the University of Waterloo and as Professor of Mathematics at New York University.

Dr. Martino served on the boards of Saint Joseph's University in Philadelphia, the World Affairs Council, the Foreign Policy Research Institute (of which he is currently a Senior Fellow), the Gregorian University Foundation, the Vatican Observatory Foundation, the Order of Malta, and numerous other boards. He currently serves on the Advisory Board of the University of Toronto's Institute of Aerospace Studies.

Dr. Martino has shared his good fortune with philanthropic works for handicapped children, merit scholarships at all levels in education, grants for cancer research, and building projects in higher education. An avid sailor, he captained his sloop "The Lady Barbara" in races in the mid-Atlantic. He served as Commodore and Board Chair of the Yacht Club of Sea Isle City, and Commodore and Secretary of the Mid-Atlantic Yacht Racing Association. He also served as the first President of the YCSIC Sailing Foundation. He is a member of the YCSIC, the Union league of Philadelphia, and the Overbrook Golf Club.

He has been honored by the Monte Jade Society, the National Italian American Foundation of Washington, and the CYO Hall of Fame in Philadelphia among others. He is also the recipient of the Saint Joseph Award from Malvern Retreat House. Dr. Martino holds honorary doctorates from Gonzaga University (Spokane, WA), Neumann University (Aston, PA) and Chestnut Hill College (Philadelphia, PA), and was knighted by Pope St. John Paul II as a Knight of Saint Gregory. He is also a Knight Grand Cross of the Equestrian Order of the Holy Sepulchre, and Knight in Obedience of the Sovereign Military Order of Malta. The Government of Canada granted him a personal Coat of Arms in 2003. In 2017 Dr. Martino was enrolled in the Wall of Distinction of the Faculty of Engineering of the University of Toronto.

Dr. Martino's lifelong accomplishments have earned him a global reputation as a scientist, inventor, financial expert, technology guru, philanthropist and author.

www.ingramcontent.com/pod-product-compliance
Lightning Source LLC
Chambersburg PA
CBHW031921190326
41519CB00007B/370